LEGENDS

OF THE EARTH

Their Geologic Origins

"Frankly, I don't take much stock in this business of the gods being angry. The way I got it doped out is that at great depths the heat caused by the enormous pressure of the earth's crust melts the stone and turns some of it into gases. See? Then you got down there a vapor tension strong enough to blow the whole works to the surface, providing, of course, a channel of weakness opens up for it."

LEGENDS

OF THE EARTH

Their Geologic Origins

DOROTHY B. VITALIANO

 *Indiana University Press*

BLOOMINGTON / LONDON

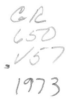

Published in Canada by Fitzhenry & Whiteside Limited, Don Mills, Ontario

Library of Congress catalog card number: 72-85855

ISBN: 0-253-14750-6

Manufactured in the United States of America

2   3   4   5   75   76   77   78

To Charles,

FOR HIS INFINITE PATIENCE

Contents /

List of Plates

List of Plates and Figures

List of Figures

List of Tables

Preface /

SINCE CHILDHOOD I HAVE BEEN FASCINATED BY mythology and folklore of all kinds, and for many years now I have been professionally involved with geology. Nevertheless it did not dawn on me that the two had a common meeting ground until in 1961 I encountered an article by A. G. Galanopoulos linking Atlantis with the Bronze Age eruption of Santorin. The term *geomythology* was conceived in November 1966 as I was describing Professor Galanopoulos' ideas to a group of colleagues. In May 1967 I found myself lecturing to the Geology Colloquium at Indiana University on the subject "Geomythology—The Impact of Geology on History and Legend, with Special Reference to Atlantis." In the audience was Professor Richard M. Dorson, director of the Folklore Institute at Indiana University, who asked to publish the talk in the journal of that institute. The enthusiastic reception of both the lecture and its slightly revised printed text, by members of the geologic profession and laymen alike, encouraged me to take seriously the suggestion by the Indiana University Press that I expand it into a full-length book.

One of the main reasons for undertaking the more ambitious enterprise has been the hope that in its own small way such a book might help bridge the communications gap between the scientist and the nonscientist. In these times, when the results of man's tampering with his natural environment are rapidly approaching crisis proportions, it is crucial that everyone, from the man in the street to those who make the laws and political decisions affecting the environment, be acquainted with the physical and biological processes that operate to create and maintain or change that environment. But so long as scientists continue to speak in a jargon comprehensible only to others in their specialized subjects, the gap between science and the humanities will continue to exist and to widen. Along with many others, I firmly believe it is our obligation as scientists to explain our subjects in terms the nonscientist can understand. If we do not, who will?

Legends of the Earth provides an offbeat framework within which to present scientific information on a variety of geologic phenomena. It cannot hope to present a comprehensive picture of our physical environment, but at least the glimpses it does afford might increase the nongeologist's understanding of some aspects of his environment and perhaps even stimulate his curiosity sufficiently that he will seek further information about the earth we live on. At the same time, the earth scientist as well as the layman should find it amusing to contrast the folklore explanations of various geologic features with the scientific explanations, and the earth scientist should be especially intrigued by the search for a possible geologic basis for certain legends and traditions. As the search has involved folklore of very diverse kinds, it has also involved a surprisingly wide variety of geologic subjects; inasmuch as geology is as highly specialized as all the modern sciences, it is possible that in these pages even the earth scientist may come across some item of information he has not heretofore encountered, in some specialty other than his own. Finally, in view of the recent proliferation of works concerning the Aegean origin of Atlantis, the earth scientist and the layman alike might be interested in a reasonably objective analysis of that idea, from the point of view of its geologic plausibility. As anyone who has been following the subject is already aware, the suggestion that the island of Santorin might have been the site of the Metropolis of Atlantis is intimately tied in with the theory of the volcanic destruction of Minoan Crete —and on this, the results of research by my husband and myself, still in progress, promise to shed a new light, as will be seen in Chapter 8.

Numerous attempts have been made to explain specific items of folklore in terms of natural phenomena. Pertinent myths and legends have sometimes been cited in geologic works, particularly those on earthquakes and volcanoes. Here and there attempts have been made to erase prevalent misconceptions concerning various geologic phenomena. To the best of my knowledge, however, no single work has heretofore attempted to demonstrate in one volume all the ways in which geology, history, and folklore can be interrelated. Most of the material presented here has been compiled from many published and a few unpublished sources, a partial list of which is given at the end of the book. The preparation of such a compendium inevitably stimulated a few original conclusions and speculations, which are offered as such in their respective contexts.

So many individuals have furnished encouragement and helpful advice, criticism, information, illustrations, and additional examples and references that to enumerate them would occupy pages, and moreover would risk offending someone by the sin of omission. A few, however, merit special mention: Dr. Sigurdur Thorarinsson of the University of Iceland, for his numerous suggestions and amplifications concerning things Icelandic, for the use of photographs, for criticism of parts of the manuscript in the early stages, and for his painstaking review of the whole first draft; Prof. Spyridon Marinatos, General Inspector of Antiquities for Greece, for the opportunity to participate in the International Scientific Congress on the Volcano of Thera in 1969, without which my information concerning the problems of the demise of Minoan civilization would be hopelessly out of date, and for his sponsorship of my husband's and my research on Crete and hospitality at the dig on Thera in the summer of 1971; Prof. Angelos G. Galanopoulos, Director of the Seismological Institute of the National Observatory in Athens, whose works, although I do not fully share his views, provided the original inspiration for *Legends of the Earth;* and Dr. Howard A. Powers, retired Director of the Volcano Observatory in Hawaii National Park, for his very enthusiastic cooperation in connection with Hawaiian volcano folklore and for his thorough review of the first draft of the manuscript.

Picture credits are included in the captions of photographs. Unless otherwise specified, drafted illustrations are the work of Mr. James R. Tolen, Senior Draftsman, Department of Geology, Indiana University.

1 /

Explanations

GEOMYTHOLOGY IS A NEW WORD. WHAT EXACTLY does it mean? As originally conceived [264],* it was defined as the geologic application of euhemerism. Euhemerus, a Sicilian philosopher of about 300 B.C., held the theory that the gods of mythology were but deified mortals; hence, euhemerism is the interpretation of myths as traditional accounts of historical personages and events. In that sense, then, geomythology seeks to explain certain specific myths and legends in terms of actual geologic events that may have been witnessed by various groups of people.

Geomythology also includes what folklorists call etiological or explanatory myths, those made up to account for various features of man's environment. Because of what has aptly been characterized as the "folklore-attracting qualities of dramatic geography" [23:97], there is a whole host of these myths and even some pseudo-folklore (or *fakelore*, as Richard M. Dorson has termed it [23:2]) purporting to explain the origin of striking landforms the world over. Even small-scale features, such as individual minerals, have occasionally inspired folklore. And needless to say, etiological myths which attempt to explain volcanic eruptions and earthquakes have proliferated in those parts of the world subjected to these sinister and spectacular recurrent geologic phenomena.

Although these are the two main ways in which geology and

* Numbers in brackets refer to the corresponding entry in the references provided at the back of this book; a second number after a colon indicates the page number in the reference.

1

folklore may be related, they are not the only ways. Popular misconceptions concerning matters geologic constitute a perfectly valid form of modern folklore. There is also the very opposite case, where what is thought to be an old wives' tale turns out to have a scientific basis after all. Finally, mythology in a very modest way has had some effect on geology, largely in matters of nomenclature but occasionally in less passive ways. All these possible relationships between geology and folklore will be illustrated in the pages to come, for exploring these relationships is the sum and substance of geomythology.

Before getting down to the business of demonstrating the ways in which geology, history, myth, legend, and folklore are interrelated, it might be a good idea to establish precisely what we mean by each of those terms, for they convey different things to different people.

If you were to ask the man in the street what he thinks is meant by geology, the chances are that he considers it to be the study of rocks. However, there is much more to it than that. The word *geology* means "science of the earth," and the earth consists not only of the solid globe made up of rocks and soils (lithosphere) and metallic core, but also of the waters on it and within it (hydrosphere) and the air above it (atmosphere); a magnetic field surrounds it (magnetosphere), and together with its moon it is part of the solar system. Its habitable land surface and waters and air constitute the biosphere, the realm of living things. All these -spheres affect one another to a greater or lesser extent, so that it is impossible to draw clean-cut lines separating geology from other physical and natural sciences—astronomy, physics, chemistry, biology—or separating the various branches of earth science from each other. These interrelationships are reflected in some of the names designating the inter- and intradisciplinary fields of earth science: for example, geochemistry and geophysics, which deal specifically with the chemistry and physics of the earth; geography (description of the earth), which is concerned with the earth primarily as man's environment; geomorphology, the study of the shaping of the earth's surface by various internal and external forces; geohydrology, concerned with water supplies, particularly those in the ground; geochronology, dealing with the absolute age of rocks; and even geopolitics, or political geography.

To these we now propose to add geomythology, which involves geology, history, archeology, and folklore—in other words, the natural sciences, social sciences, and humanities. Thus geomythology is undoubtedly the most interdisciplinary geoscience of them all.

The next question is, what exactly do we mean by *myth, legend,* and *folklore?* The answer is again anything but simple, for the distinction between myths and legends has been and still is the subject of much discussion and disagreement among those concerned with such matters. H. J. Rose [211:6] prefers *legend* as the all-embracing term, and subdivides all traditional tales into *myths* proper, which are the "result of imaginative reflection"; *sagas,* which have an historial basis (and which he acknowledges are also known as legends); and *märchen,* tales intended purely to amuse. J. H. Brunvand [23:1], on the other hand, uses *folklore* as the general term, defining it as "the unrecorded traditions of a people," that part of a culture that is passed on orally. Folklore includes myths, legends, and folktales; both myths and legends may be regarded as true by their tellers (in contrast to folktales, or *märchen*), and are distinguished primarily by "the attitude of storytellers to them, the settings described in them, and their principal characters. Myths are regarded as sacred, and legends as either sacred or secular; myths are set in the remote past in the otherworld or an earlier world, and legends in the historical past; myths have as their principal characters gods or animals, while legends generally have humans in the major roles" [23:79]. For geomythological purposes it is most convenient to consider as myths those traditions which are purely etiological, and as legends those which are actually euhemeristic, following Brunvand to the extent of using folklore as the overall term. Strictly speaking, therefore, the subject of this book should be termed *geofolklore;* but inasmuch as a clear distinction between myths and legends is impossible in any definition, and the terms have come to be used more or less interchangeably, we will use *geomythology* in a broad sense, referring to any geologically inspired folklore regardless of its origin.

Many theories have been advanced to explain the origin of myths, each theory supported vigorously by its adherents, and all too often to the exclusion of any other. Fundamentally these theories have sought to account for the notable similarity of elements in different bodies of mythology, and for similarities between myths and folktales. Such similarities can be explained in either of two basic ways: by independent invention in different places (polygenesis), or by transmission of a single invention to other regions (diffusion). Diffusion can also lead to what folklorists call syncretism, whereby elements of separately evolved traditions become blended into one. Modern comprehensive theories recognize that the myths, legends, and folktales of any nation most probably have sprung from a combination of sources rather than from any single source. Although geomythology is concerned only with that small segment of the vast body of folklore which is related to the physical environment, it should be apparent in the pages to come that geomythology amply demonstrates the validity of this multiple-origin viewpoint.

Next we come to history. In the strict sense of the word, *history* means only what is preserved in written records. Myths and legends usually date from long before the use of writing by the culture concerned, so whatever factual basis they may have would be prehistoric according to this definition. Moreover, as writing developed at different times in different places, what is prehistoric in one place may be contemporaneous with what is truly historical in another. In this book we will be using *history* in a very loose sense, to designate not only what is known from written sources, but also traditions which were originally oral (semi-historical traditions) and, more important still for our purposes, all that is "written" in the geologic and archeological records, from which we can deduce the possible facts underlying a legend. It will be clear as we go along that the farther back we go in time, the fuzzier becomes the line dividing history (that is, what really happened) from legend. Thus a word about the relative reliability of oral tradition, of historical records, and of scientific evidence is in order at this point.

The reliability of oral traditional history has been the subject of raging controversy since the time of Euhemerus, if not before [52]. The extreme skeptics deny that time-worn traditions contain any

4

scrap of historical truth whatever; the champions of oral tradition, while not denying that fictional elements are present, claim it is possible to separate the historical from the nonhistorical. If we think in terms of our modern world, in which a piece of gossip may be relayed with substantial changes at every retelling until all semblance of truth is lost, it is difficult to believe that any resemblance to fact could long survive in oral tradition. But there is another modern analogy which might be more apposite. How often, parents, have you retold a bedtime story to your offspring and found yourself corrected if you deviated by so much as a word from the first version the youngsters had heard? In the very same way, would it not be difficult to introduce variations into the tales told by professional storytellers to peoples whose cultures remained virtually unchanged for thousands of years—people, for instance, like the American Indians before the intrusion of the paleface, who depended on storytellers to amuse them and to preserve the tribal traditions? Under such conditions changes would creep in only with exceeding slowness. When history has been handed down through a succession of professional storytellers with highly trained memories, like the Icelandic skalds [95:52] for instance, there is every reason to believe that when put in writing after a few hundred years it might still be reasonably accurate. However, it must be agreed that the first written version of earlier events inevitably would differ to some extent from the original facts.

Written history is of course much less subject to change than oral tradition, but the written word is ultimately only as reliable as its original chronicler. There is always the chance that the information was already distorted before it reached its recorder, or that even a first-hand account may be inaccurate. Consider the all-too-frequent discrepancies between one's own recollection of some event and the newspaper account of it next day. There is also the well known fact that ten eyewitnesses may give up to ten different versions of an incident, particularly if it was brief and startling.

Even after an account is written down, departures from the original can be introduced later by careless copying (in the days before printing) or by faulty translation. Translators are not infallible. The unique and forceful imagery of the New Testament proverb "It is easier for a camel to go through the eye of a needle, than for a rich man to enter the kingdom of God" is thought by some to stem from a translation error in which someone mistook the Greek word καμηλος, meaning rope or hawser, for καμελος, meaning camel.

Apart from the question of a simple mistake in translation, there is the much more subtle question of the precise rendition of meaning; this often involves a degree of interpretation and may lead equally well qualified translators to come up with different versions of the same material. A frivolous example comes to mind: Some years ago there was a popular ditty entitled "The Purple People Eater." If one listened to the words it was made clear in the punch line that the beast in question was not a fearsome purple monster which devoured any hapless people who fell into its clutches, but a friendly and much misunderstood creature dangerous only to purple people. Now suppose one were trying to translate that title into French, not knowing which meaning was intended. Should it be "le mangeur de gens pourpres" (the eater of purple people) or "le mangeur-de-gens pourpre" (the purple eater of people)? Only one choice would be correct, while either choice (in print, at least) would lose the ambiguity which was the whole point of the song. In the vast majority of cases the context indicates the correct choice of word in translating; but there are cases where it does not, and "something has been lost in translation" is no idle phrase.

Finally, there can be deliberate distortion introduced at any time into the historical record to suit some devious purpose or to fit in with some prevailing philosophy—Hitler's "Aryan" philosophy, for instance, or dialectical materialism.

In general, the geologic and archeological records are not subject to deliberate falsification. Of course, to the extent that it is a conscious human record the archeological record may include some distortions on the part of long-gone people, such as the glorification of quite undeserving rulers in painting and sculpture; but due allowance can be made for that. Real scientific hoaxes are so rare that if they succeed in fooling just a few people for just a little while, they make history. In geology there have been only two instances of such deception which came anywhere near succeeding. One of these, the Beringer case, was perpetrated at the expense of a single individual rather the science as a whole [122]. The other hoax, the Piltdown forgery [267], came somewhat nearer to being the "perfect crime," but even that fooled only some of the people some of the time.*

One might well ask whether there could have been hoaxes so clever that they have never been detected and never will be. However, to find in one person both the desire to fool scientists and enough scientific knowledge to plan and execute a successful fraud (as appears to have been the case in the Piltdown forgery) is *extremely* rare; to find someone with the necessary expertise who would be willing to lend himself to a conspiracy involving several individuals would be even more difficult; and finally, to find that the fabricated "evidence" fitted so well with true later discoveries that it did not eventually stand out in all its falseness, as Piltdown Man did, would be an inconceivable coincidence—and in such a case, what harm would have really been done?

But while the geologic and archeological evidence in itself is very unlikely to lie, it is so fragmentary that our interpretation of it may be quite erroneous. The situation is analogous to a gigantic jigsaw puzzle, of which we are given just a few scattered pieces at a time. We have to try to fit them into an overall pattern, but until such time as we have enough pieces that the design is unmistakable, we must link them together with a network of pure conjecture, which is either reinforced or torn to shreds as new pieces of the puzzle become available.

Scientific theories are analogous to etiological myths insofar as they both are attempts to explain observed facts. But whereas myths are the product of the naive imagination and call upon the supernatural or the physically impossible to explain those facts, scientific theories must be consistent with everything that is known about the natural world up to their time. Myths die hard, even when incontrovertible facts are marshalled against them. Scientific theories also die hard, their most zealous adherents clinging to them so long as there is any faint possibility that the pieces of the jigsaw might fit as they would have them fit. But once its underlying assumptions are proved untenable, a scientific theory must be rejected by anyone claiming to be a scientist.

* A summary of both these cases will be found in Appendix A.

2 /

Geology's Role in History

and Legend

BY FAR THE GREATER PART OF ALL GEOLOGICALLY inspired folklore can be classified into the two main categories mentioned earlier, etiological and euhemeristic. Except for those which explain earthquakes and volcanism, etiological geomyths nearly always concern geologic features which were formed long ago. For, like the mills of the gods, geologic processes as a rule grind exceedingly slowly, requiring thousands to millions of years to unfold. The vertical uplift or subsidence of substantial areas of the earth's crust, the crumpling of mobile crustal belts into folded mountains, erosion and deposition of sediments, volcanism, continental glaciation—all these have been going on either continually or intermittently since the beginning of geologic time, more than three and a half billion years ago. Most of the action—if that word can be applied to such slow-motion processes—was over and done with long before man existed. The first "modern" man, Cro-Magnon Man (*Homo sapiens sapiens*), appeared about fifty thousand years ago; and if Neanderthal Man is admitted to our ranks (as *Homo sapiens neanderthalensis*), then our species, as distinct from various primitive, barely human creatures who date all the way back to more than three million years ago, has been around for only one hundred thousand years at most—the merest blink of an eye when compared to the whole duration of geologic time (see Table I).

TABLE I
The Geologic Time Scale

Era	Period	Epoch*			Age (Years)
Cenozoic	Quaternary**	Holocene (Recent)			10,000-11,000
		Pleistocene			1,000,000
	Tertiary**	Neogene	Pliocene		13,000,000
			Miocene		25,000,000
		Paleogene	Oligocene		36,000,000
			Eocene		58,000,000
			Paleocene		63,000,000
Mesozoic	Cretaceous				135,000,000
	Jurassic				181,000,000
	Triassic				230,000,000
Paleozoic	Permian				280,000,000
	Carboniferous	Pennsylvanian***			310,000,000
		Mississippian***			345,000,000
	Devonian				405,000,000
	Silurian				425,000,000
	Ordovician				500,000,000
	Cambrian				600,000,000
Precambrian	Proterozoic				2,700,000,000
	Archeozoic				3,500,000,000

PRIMORDIAL EARTH

Age of the Earth—4,500,000,000

* Subdivisions are given only for the Cenozoic.
** The Paleozoic and Mesozoic were originally called the Primary and Secondary, respectively.
*** The Carboniferous is regarded as two systems in North America only.

The very slow geologic processes have had an equally important effect on history and on myth. These slow processes determined the present distribution of lands and seas, the topography of the land surface, the nature of the rocks underlying that surface, the distribution of mineral wealth and water supplies, and all other such aspects of our planet. These factors in turn determine which regions are most desirable to inhabit, which nations are the "haves" and which the "have-nots," which areas are easily accessible, which regions are most vulnerable to attack and which most defensible, and so on and on. Many books have been written about political geography, or about the role of minerals in world history, but such subjects are beyond the scope of geomythology. We are concerned primarily with the superabundant folklore inspired by these geologic processes.

Every nation has its folklore accounting for landscape features; rarely, however, and then mainly by coincidence, do etiological myths bear the slightest resemblance to the real geologic history of the features which inspired them. Myths of this type might be called *ex post facto* geomyths, for they are made up to explain the end results of processes whose action was not witnessed. Chapter 4 is devoted entirely to such folklore, and other examples will be found in the more specialized contexts of volcano folklore and deluge traditions.

The very long-term geologic processes are the sum total of an infinite number of tiny individual events, usually too small to be noticed even though they are going on around us all the time. Occasionally they add up fast enough, in some particular setting, for their results to be perceptible over a human lifetime or over several generations. Processes of this type include changes in the relative level of land and sea, the advance or retreat of glaciers, climatic changes, and the silting up of harbors. As will be shown in Chapter 3, these have had considerable effect on the migration of very early man, but in more recent times their overall impact on human history has been relatively minor and their effect on legend almost nil.

Occasionally a single event in the overall geologic picture occurs very rapidly and affords us a glimpse of the tremendous forces operating within the earth, compared to which the most formidable of man's efforts to date, the hydrogen bomb, is no more powerful than an oversized firecracker. Great earthquakes, volcanic eruptions, floods, landslides—these can kill people or drastically alter the course of their lives. When many are involved, whole communities may be disrupted temporarily or permanently, and in a very few notable cases such events have even significantly influenced the affairs of whole nations. In general, however, the impact of these rapidly occurring events on human history has been negligible compared to the influence of the very slow geologic processes which shaped the environment. Yet these events are the stuff of which legends are made—euhemeristic legends in which a germ of truth is enveloped (like the grain of sand in an oyster which initiates the growth of a pearl) in layer upon layer of shiny fiction. Abundant examples will be found in Chapters 6 through 10.

But why, if their impact on human history has usually not been great, should it so often have been this type of event which engendered legends? I believe it is because at the time they happened they were eminently *newsworthy*. For unless human nature has changed considerably through the ages, what is considered news, and therefore what may be remembered when the normal events of daily life are long forgotten, is the unusual, particularly the violently unusual. And what is more violently unusual than a natural catastrophe? In order to understand how geologic events could have inspired legends, let us look at a few instances from modern, or at least historical, time when geologic forces have altered the normal course of human affairs. Examples are so numerous that some criterion had to be established to guide the choice of those to be included, and accordingly I have selected those which illustrate the point that the concept of catastrophe is very highly subjective— in other words, the magnitude of a disaster caused by geologic forces is by no means a measure of the magnitude of those forces, but only of their "newsworthiness." Furthermore, I have tried to avoid instances which have already been written up time and time again, except where they are particularly relevant.

On May 8, 1902, an eruption of the type called a "glowing cloud" swept down the flanks of Mont Pelée on Martinique, annihilating all but two [25:109] of the more than thirty thousand inhabitants of Saint Pierre at the foot of the volcano. (One of the survivors, ironically, was a condemned murderer in a dungeon.) I seriously doubt that the eruption would have been remembered at all, except perhaps by volcanologists, had the fiery cloud taken a different route down the mountainside and spared the city. In terms of volcanic energy the Pelée eruption was no more remarkable than many others in our time. In 1915 there was a very similar blast from Lassen Peak [121:111] in California, but there nothing was harmed other than several thousand acres of forest and whatever wild life inhabited it. Three of the truly most violent eruptions in all history have occurred in this century—Katmai in Alaska in 1912 [88], Bezymianny on Kamchatka in 1956 [85], and Sheveluch on Kamchatka in 1964 [86]; but all of them occurred in uninhabited regions, and they are of interest only to volcanologists and to the few who might have witnessed them or felt their indirect effects from afar. The Pelée eruption, however, is popularly thought of as one of the great eruptions of all time; actually it was one of the great *catastrophes* of all time, which is not quite the same thing.

The complete subjectivity which governs the reporting of natural catastrophes is aptly illustrated by the Nevada earthquakes of 1954 [27, 247]. On July 7 of that year the Washington *Post* reported, in a small article on the front page, that an earthquake had shaken the area around Fallon, Nevada, during the night, causing some damage to buildings downtown and seriously disrupting the irrigation system; it also mentioned that a second shock had occurred in the afternoon. In San Francisco, the *Examiner* carried a front-page headline and pictures of the damage in the Fallon area, but on the whole they were much more interested in the fact that the second shock had been felt in San Francisco. In Nevada, particularly in the Fallon area, the shocks were of course a major concern.

Fallon, an agricultural center on the edge of Carson Sink, depends on water from the Carson River, and therefore the damage to the irrigation system was a very serious matter. The big question in everyone's mind, particularly after the second strong shock, was, "Will it happen again?" A temporary seismograph installed in the basement of the City Hall recorded numerous aftershocks daily; few were perceptible except to the instrument, but they were very many and indicated that stresses in the earth were still being released. But all except the local residents forgot about Fallon soon after the shocks—until on August 23 another strong shock, centered a little more to the east at Stillwater, undid all the temporary repairs just completed on the irrigation ditches [225]. That shock did not rate a single line in the Washington *Post*, despite the fact that its magnitude was exactly the same as that of the July 7 shocks (6.8). The *Examiner* merely reported that the shock had caused some damage again in Fallon and Lovelock, and had been felt in

San Francisco as well as in Boise and Salt Lake City. And in Fallon, aftershocks continued to be a cause for concern.

The series of earthquakes culminated with two more strong shocks on December 16, again centered to the east of the last one, in Dixie Valley and near Fairview Peak [226]. (The Dixie Valley shock was accompanied by spectacular surface faulting which can be seen if one makes a short detour off U.S. 50 where indicated by a sign.) Strangely enough, the only dollar damage was that done in Sacramento, 185 miles away as the crow flies, and all of that was caused by the surging of fluids in tanks—at the municipal sewage treatment plant, in a covered reservoir, and at a soup company's clarifier tank. Because of this damage, the new shocks rated a small mention in the Washington *Post*. The *Examiner* had a front page article—under the headline: "Two new sharp quakes jolt San Francisco Bay area"—which dwelt mainly on the effects of the shocks as perceived in San Francisco, which were exceedingly minor. Passing mention was made of the damage in Sacramento, but there was not a word about Nevada, where the earthquake occurred!

It should be obvious from this example that the newsworthiness of any disaster caused by geologic forces depends primarily on human interest, which varies with the extent of the damage and distance from the scene. The greater the loss in lives and property, the wider the interest. The greatest of geologic upheavals is not news if it happens where there are few or none to suffer the consequences, whereas a relatively small upheaval in a densely populated area, such as the earthquake near Los Angeles on February 9, 1971 (which was not even as strong as the Nevada shocks in 1954), can cause much suffering and make headlines around the world.

Occasionally, however, a small geologic event in an out-of-the-way place does make headlines even though no blood is spilled. Such was the 1961 eruption of Tristan da Cunha [8]. The island of Tristan da Cunha is the top of a volcanic peak on the outer slope of the Mid-Atlantic Ridge near its southern end (Fig. 1). The inhabitants of the island are the descendants of British soldiers stationed there at the time Napoleon was exiled to Saint Helena, another peak on the Mid-Atlantic Ridge, or of whalers who settled there. When an erutpion took place from a new vent only three hundred yards from Settlement, where the entire population lived

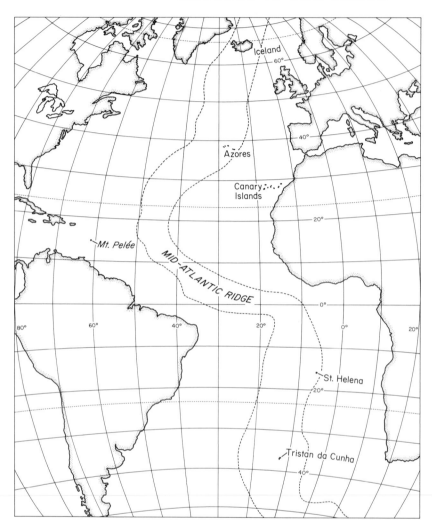

FIGURE 1. *Sketch map of the Atlantic Ocean, showing the Mid-Atlantic Ridge and other places mentioned in this and subsequent chapters.*

(all 264 islanders plus 11 officials with their families), everyone had to be evacuated to England.* As eruptions go, it was insig-

* As this book goes to press, an eruption on the island of Heimaey in the Westman Islands off the southwest coast of Iceland is very similarly attracting worldwide attention by forcing the evacuation of the prosperous fishing town of Vestmannaeyjar.

nificant. Lava covered less than a quarter of a square mile and ash half as much. Nobody was hurt. The only living things that suffered were plants growing near the newly built cone, poisoned by toxic fumes; the animal life was not noticeably affected. No doubt the inhabitants would have been able to remain throughout the eruption, had the damage not been concentrated on Settlement Plain, and had it not included destruction of the cannery, which was an integral part of the one and only industry, fishing. As it happened, the whole life of the tiny, isolated, inbred community was disrupted. Even this disruption proved to be temporary, however. It is interesting that most of the islanders could not get used to the bewildering modern world into which they were thrust and scattered, and almost all, young and old, chose to return to the quiet, close-knit, relatively primitive life of the island when the danger was over.

Another disaster which made headlines around the world was the Tangiwai disaster in New Zealand. Through a freakish combination of circumstances, what might have been no more serious than an interruption of railroad service became an accident that claimed 151 lives on Christmas Eve in 1953. And, although it was quiescently minding its own business at the time, the volcano Ruapehu was the ultimate culprit [171]. The Whangaehu River has its source in a glacier high up on the 9,175-foot volcano, the highest peak on New Zealand's North Island (Fig. 2). In its last eruption in 1945, a lake had been formed in the crater, dammed up by a combination of lava, ash, and névé. (Névé is granular snow at the head of a glacier, not yet compacted into ice.) Melted by heat from the volcano and also no doubt from the summer sun (New Zealand being in the Antipodes), the névé portion of the dam suddenly gave way and released a flood which roared downstream, picking up material to form a mud-and-boulder flow of the type called a *lahar*.* The railroad bridge twenty-five miles downstream near Tangiwai was demolished when the lahar struck it, and for those aboard the approaching express the timing was diabolical.

* *Lahar* is a Javanese word used in volcanology to designate a torrential volcanic mud flow, caused by the bursting of a crater lake as in this case, or by the melting of snow and ice by volcanic heat, or by the washing down of ash resting on steep volcanic slopes in heavy tropical rains.

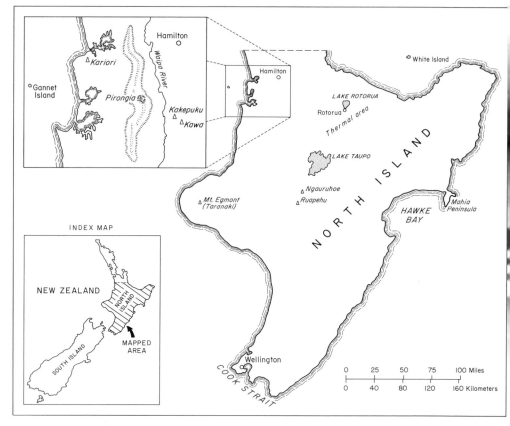

FIGURE 2. *The North Island of New Zealand, showing places mentioned in various Maori legends.*

Had the bridge gone out a few minutes later the train would have been safely over; minutes earlier, and the train might have stopped in time. But as it was, several cars plunged into the gorge, causing great loss of life.

Although the Tangiwai disaster cast a pall of gloom over Christmas in New Zealand that year, it did not in the long run affect anyone other than the victims and those close to them. Even disasters of the magnitude of the San Francisco earthquake of 1906 or the Tokyo earthquake of 1923, staggering in their toll of lives and property, were weathered by their respective communities and countries. But at least one geologic disaster in historical time did nearly ruin a whole country, the Lakagígar ("Laki craters") eruption in Iceland [240].

Iceland is one of the most active volcanic areas in the world. The whole island is essentially a pile of lavas, built up during and since Tertiary times on the Mid-Atlantic Ridge near its northern end (see Fig. 1). In the greatest lava eruption in historical time, lava poured without ceasing for eight long months, from vents along the 25-kilometer-long Laki fissure (Fig. 3). The eruption began on June 8, 1783. Lava found its way down two river valleys to the plains, where it spread out and ultimately covered a total area of 565 square kilometers with an estimated volume of 12 to 15 cubic kilometers of lava, now known as the *Eldhraun* ("fire flow"). Today the main road runs for mile after mile across a weird landscape which resembles the setting of some science-fiction story laid on another planet, especially when the backdrop of ice-

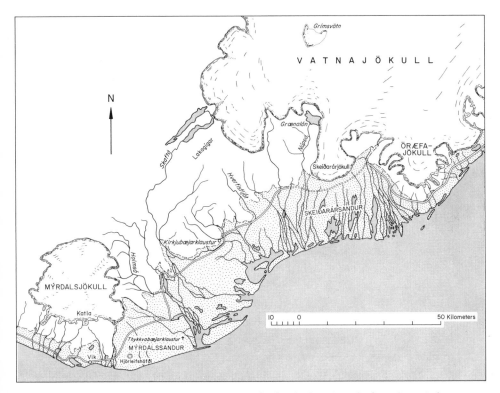

FIGURE 3. *Sketch map of southeastern Iceland, showing the location of the Lakagígar crater row, the subglacial volcano Katla, and other features mentioned in this and subsequent chapters.*

17

covered mountains is obscured by mist. The thick cover of moss and lichen already developed on the lava in the rainy climate intensifies this impression (Plate 1). Although a number of farms

PLATE 1. *Eastern branch of the Laki lava flow of 1783, which almost wiped out Iceland. View toward the east. The ice-capped peak in the right background is Öraefajökull, the highest mountain in Iceland; it is a formidable active volcano, topped by a glacier. (Photo by Sigurdur Thorarinsson.)*

were lost under the lava, the greatest harm was done not by the lava itself but by the "bluish haze" accompanying it. This haze contained noxious gases, particularly sulfur dioxide, and it hung over most of Iceland all that summer and stunted the grass crop. This was a disaster of unparalleled magnitude, for on the hay crop depended the livestock, and on the livestock, the people. In the notorious "haze famine" which ensued, 77 percent of the sheep, 76 percent of the sturdy Icelandic ponies (essential for transportation in those days), 50 percent of the cattle, and 20 percent of the people died from hunger or, weakened by malnutrition, succumbed to disease. In the light of experience in the Hekla eruptions of 1947-48 [238] and 1970 [239], it is now believed that fluorine in the volcanic ash erupted along with the lava was partly responsible for the loss of livestock [240]. It takes a hardy nation—which the Icelanders most certainly are—to survive such a blow.

Another Icelandic eruption of a more run-of-the-mill kind helped influence a very important decision in that country's history. To appreciate the story fully, it is necessary to understand the geologic setting. When the first settlers arrived from Norway in about A.D. 874, they were struck by the contrast between the fresh post-glacial lavas and the older rocks. A trained geologist today has no difficulty in recognizing that the older rocks also were lavas (Plate 2), but we must keep in mind that even in the early

PLATE 2. *Preglacial lavas exposed in the gorge below Dettifoss, Iceland's highest waterfall. (Photo by the author, August 1960.)*

nineteenth century, when geology was in its infancy, a great controversy raged over the origin of extensive fine-grained, layered rocks of this type. Adherents of the "neptunist" point of view claimed these rocks were sediments deposited in the sea, like sandstones, shales, and limestones; the "plutonists" insisted these were volcanic, crystallized from lavas poured out on the surface like modern lavas. It was not until 1858, when a technique was developed for making very thin slices of rocks to study under the polarizing microscope, that the question was unequivocally settled in favor of the plutonists. The postglacial lavas (Plate 3), on the other hand, were unmistakably the products of volcanism. There is a spe-

PLATE 3. *Post-glacial lava* (hraun), *Herdubreidarlindir, Iceland. This flow exhibits the ropy surface of the type of lava called* pahoehoe, *and also a pressure ridge, opened up when the chilled crust of the still-moving flow was compressed. (Photo by Claude M. Roberts, August 1972.)*

cial word in Icelandic for these recent flows, *hraun* (pronounced "hroin"), which can best be translated as "lava field." Anyone who has seen any of the lava beds in the western United States has seen *hraun;* notable examples are those in Lava Beds National Monument in California, in Craters of the Moon National Monument in Idaho, in Sunset Crater National Monument in Arizona, and along Highway 66 near Grants, New Mexico.

In A.D. 930 Iceland established the Althing, the first and most continuous democratic parliament in the world. Representatives from all parts of the country met every summer in Thingvellir (Plate 4) where a natural cleft, the Almannagjá, provides excellent acoustics for a speaker standing on Law Rock. In A.D. 1000 the main

PLATE 4. *The Almannagjá at Thingvellir, Iceland. The world's oldest democratic parliament, the Althing, used to meet in the open air at this natural volcanic cleft. The flagpole marks Law Rock, where the speaker stood. (Photo by the author, August 1960.)*

item on the agenda of the Althing was the burning question whether Iceland should officially embrace Christianity or continue to worship the old Norse gods. Skillful arguments were presented on both sides, and no decision could be reached. According to the Kristnesaga, the official history of Christianity in Iceland, the debate had reached the point where "each man declared the other outlaw, the Christian men and the heathen men against one another" [258:399]. Just at that point a messenger came bearing the news that lava was pouring from a fissure in Aulfus, about twenty miles east of Reykjavík, and that it threatened the farmstead of the chieftain Thorodd (Fig. 4). "Then the heathen men began to say, 'It is no wonder that the gods are wroth at such speeches.'" The tide of opinion threatened to turn in their favor, but the chieftain Snorre, a

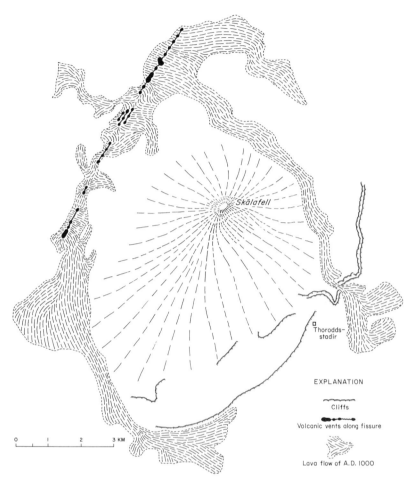

FIGURE 4. *The lava flow of* A.D. *1000 near Hveragerdi, which played a decisive role in the history of Christianity in Iceland. Lava issuing from vents along a fissure flowed around the hill Skálafell, threatening the farmstead of the chieftain Thorodd at its foot. (After Einarsson, 1960* [55]*.)*

spokesman for Christianity, seized the opportunity and turned the news to his own advantage. Gazing out over the vast expanse of *hraun* which floors the whole valley at Thingvellir, he asked, "What were the gods wroth over then, when the lava on which we now stand was burning here?" [258:400]. The vote ultimately went in favor of Christianity.

Oddly enough, a volcano also played an important part in the acceptance of Christianity by another people at the opposite end of the world from Iceland. That is the story of Kapiolani [270:152], immortalized by Tennyson in a poem bearing her name. Kapiolani, wife of the high chief of the Kona district on the island of Hawaii, was an early convert to Christianity. In December 1824, when she had been a convert for three years, she resolved to show her people that Pele was not a deity to be worshipped and propitiated, but merely a heathen superstition. Still under the influence of that superstition, her husband and friends tried to dissuade her, and the priests and priestesses of Pele prophesied a horrible doom. It was more than a hundred miles from her home in the western part of the island to the crater of Kilauea, where Pele and her numerous relatives were believed to dwell. The journey was arduous, much of it on foot and sometimes across the most rugged kind of "aa"* lava terrain. To those along the way who begged her not to go on, she replied, "If I am destroyed, then you may all believe in Pele, but if I am not, you must all turn to the true writings" [270:159]. Upon reaching the Pit of Pele she broke several taboos: she picked and ate sacred *ohelo* berries without first offering some to Pele, she threw stones into the crater, and she descended several hundred feet to stand at the edge of the lava lake (Plate 5). When she returned unharmed and the people saw that she had come to no dreadful end, they realized that Pele was not as much to be feared as they had believed. In the ensuing years there were many converts throughout Hawaii, among them the High Priest of the volcano.

---

* Two Hawaiian words have been adopted internationally by volcanologists to designate the main types of lava field—*aa* and *pahoehoe*. Pahoehoe (pronounced pa-ho'-ee-ho'-ee) has a smooth, billowy, or ropy surface (see Plate 3) and is formed from more fluid lava than aa. Aa (pronounced ah'-ah) consists of a jumble of clinkers, some of them incredibly rough and jagged (see Plate 1). Aa terrain is fiendishly difficult to traverse. The Hawaiians of old used to lay smooth water-worn boulders, laboriously hauled from the shore, to form trails of stepping-stones across the otherwise virtually impassible aa. I have been told on unimpeachable authority that some real estate ads in West Coast newspapers have been known to describe certain Hawaiian land offerings as "good AA land"—which without prevarication nevertheless manages to convey the impression that this land is even better than Grade A. Caveat emptor!

PLATE 5. *Kapiolani defying Pele at Kilauea, 1824. From a painting by Peter Hurd. (Courtesy of Amfac, Inc., Honolulu, Hawaii.)*

In the Philippines nearly two and a half centuries before Kapiolani's demonstration of faith, a Franciscan monk named Estaban Solis performed a similar service in the cause of Christianity. In 1592 he ascended the dread volcano Mayon to persuade the superstitious natives that no evil spirit resided in the mountain. He was driven back short of the summit by choking gases and vapors emanating from the volcano, but the natives were convinced and allowed themselves to be baptized. The monk was not as fortunate as Kapiolani, however; within a year he sickened and died, presumably from the effects of the poisonous gases he had inhaled.

Some forty years before Kapiolani, Kilauea itself had directly intervened in the affairs of Hawaii. The last chief to contest Kamehameha's right to rule the "Big Island" was his cousin Keoua [270:139]. In November 1790, Keoua and his army, accompanied by their women and children, were camped near the crater. During the night the volcano erupted, and fearing that they had offended Pele by rolling stones into the crater, they spent the whole next day vainly trying to appease her. On the morning of the third day they started on their way, divided into three companies. Just as the central party had passed the crater, the earth shook and a tremendous blast of steam, burning ash, and rocks burst forth and annihilated every one of the group. Even a few of the forward party were killed or injured, but although the rear party was actually closest to the crater at the time, it was not in the direct line of Kilauea's fire and escaped unharmed. As violent explosive activity is extremely rare in connection with Kilauea eruptions [108:77], the event was hailed as a sign that Kamehameha was under Pele's special protection. Keoua, in the forward party, had been spared and continued his fight, but Kamehameha prevailed and eventually all the islands were united under his rule [108:33].

About the same time that the history of Christianity in Iceland was being influenced by an eruption, the flourishing Hindu-Javanese kingdom of Mataram in central Java suddenly vanished after a brief decline. Its disappearance was one of the great mysteries of Javanese history. The center of power and culture shifted to eastern Java, and central Java was not heard from until the second State of Mataram, founded by Mohammedans, rose to power after 1595. The Calcutta Stone, so called because it was brought to Calcutta from Java by Sir Thomas Stamford Raffles (Lieutenant Governor of Java, 1811-1816), bears an inscription commemorating the founding of a hermitage by King Erlangga in East Java in A.D. 1041. This inscription describes the destruction of the old kingdom of Mataram in the year 928 Sjaaka (A.D. 1006). Historians could not agree to the exact meaning of the account; some thought it symbolically referred to wars or pestilence, others took it more literally as a reference to a great natural catastrophe. The Dutch geologist R. W. van Bemmelen [10, 11] made a thorough study of the history of the volcano Merapi and of the coastal region near Semarang, and concluded that the old State of Mataram had been a victim of not one, but two blows dealt by geologic agencies. He found that there had once been a fine natural harbor on the north coast of Bergota, near the present town of Semarang. The prosperity and cultural importance of Old Mataram depended on that harbor, for there were no safe harbors on the south coast. By the beginning of the tenth century the silt carried down by the river began to fill the harbor at Bergota, and some time between 916 and

927 it was abandoned as a port. Then in 1006 Merapi dealt the coup de grâce.

In the inscription telling of the destruction of Old Mataram it is hard to sort fact from fiction, but the volcanological record is clear. It shows that the old cone of Merapi collapsed in a cataclysmic fissure eruption which spread a thick blanket of ash over central Java, undoubtedly destroying its fertility for many decades and completely disrupting the drainage pattern. Using old maps and descriptions of the amounts of matter erupted in the period 1833 to 1942-43, available because Merapi had been under scientific observation for more than 150 years, van Bemmelen estimated that it had taken just about nine and a half centuries for the new cone of Merapi to have been built up on the ruins of the old—just as Vesuvius has been built up since A.D. 79 on the ruins of Monte Somma. Nine and a half centuries is close enough to A.D. 1006 to suggest very strongly that the eruption was indeed the natural catastrophe referred to in the Calcutta Stone inscription, and that it was nature, not war, that destroyed Old Mataram.

From Old Mataram, poised on the threshold separating history from legend, it is an easy step back into the realm of euhemeristic folklore. Tracing such legends back to a possible geologic source is by far the most exciting aspect of geomythology, and therefore, on the principle that one reserves the best till last, we will first dispose of the other kinds.

3 /

Slow Motion

IN CHAPTER 2 WE MENTIONED IN PASSING THAT THE geologic processes whose operations are barely perceptible during a human lifetime or over a few generations have had little impact on recent history and even less on legend. In fact, they have had so little effect on the latter that they afford negative proof of the contention that only catastrophes breed euhemeristic legends.

These changes, fast in geologic terms but slow in human, include sea level fluctuations, glaciation, climatic changes, and sedimentation. The first two are largely interdependent. The exact position of any oceanic coastline depends on one or both of two factors: uplift or subsidence of the land itself, and rise or fall of sea level. The absolute position of sea level depends on the amount of water locked up in continental ice caps and mountain glaciers throughout the world at a given time. "Ice ages" are known to have occurred at certain times in earth history. During the Pleistocene period (see Table I), about which we naturally know quite a bit more because the geologic evidence is still so fresh, the ice front advanced outward from centers of accumulation in northern Canada and Scandinavia, and subsequently retreated,* at least four times.

---

* That is, the position of the ice *front* shifted back toward the north, because the rate of melting was greater than the rate of forward motion of the ice; the ice itself never actually reversed its direction of motion, which was always outward from the center of accumulation.

So little geologic time has elapsed since the last ice sheets retreated that we really do not know whether the Ice Age has ended or whether we are merely living in another interglacial age. Remnants of the northern hemisphere's ice caps still linger on Greenland and Iceland and some of the Arctic islands. It is not yet certain whether the maximums of ice advance in the southern hemisphere alternated or more or less coincided with those in the northern, but at present the Antarctic ice cap is receding too. In any case, sea level was at its lowest when the ice was at its maximum extent, and vice versa. Uplifted marine terraces in various parts of the world indicate that at its highest, sea level stood about one hundred feet or so above its present position.

The glaciers also affected the level of the solid land. Under the tremendous load of ice the rigid crust of the earth was pressed down into the underlying less rigid mantle (Fig. 5). When the load was removed, the mantle readjusted slowly and the land began to rise, and it is still rising a little. This kind of adjustment is called *isostatic*, while the changes of sea level independent of changes of land level are called *eustatic*, or *glacioeustatic*. There are various kinds of evidence that tell us the extent and rate of isostatic uplift. Raised beaches, tilted up toward the centers of glaciation, can be traced on the coasts of Scandinavia and North America, and around the Great Lakes; these beaches can be dated by fossil shells or pollens or by the radiocarbon method.* Present uplift can be measured by means of very precise geodetic leveling surveys, repeated at intervals of a decade or so, and in some cases from historical records. In Norway there has been a recent average rise of 1.2 millimeters per year at the coast in the Trondheim area and 4.8 millimeters per year at the Swedish border, a rise of about 1 meter at Bergen since A.D. 1150 and about 2 meters at Karmøy Island since A.D. 900 [134].

The Pleistocene epoch was the time during which prehistoric man evolved from an ape-like hominid to the species *Homo sapiens*. The dispersal of prehistoric man to various parts of the world, and particularly to the western hemisphere, was very intimately bound up with the world-wide glacioeustatic sea level fluctuations. At times when sea level was lower than it is today, Bering

---

* See Appendix B for an explanation of radioactive dating in general and of the radiocarbon method in particular.

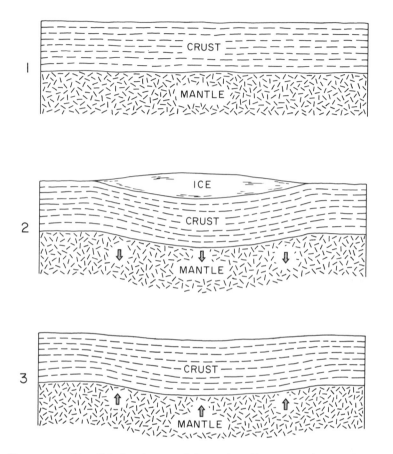

FIGURE 5. *Simplified scheme of isostatic adjustment during and after glaciation.*

1. Isostatic equilibrium prevails.

2. The earth's crust is slowly depressed under the weight of continental ice sheets.

3. After the ice load has been removed, the crust rises slowly. Equilibrium has not yet been restored, as indicated by measurable uplift still going on.

Strait was dry land and Stone Age man crossed freely from Asia and spread southward to the tip of South America and eastward to the Atlantic Ocean. The first wave of migration must have been more than 15,500 years ago at least, for artifacts that old have recently been found in the La Brea tar pits in Los Angeles, the oldest yet discovered in the western hemisphere. A few individuals might have accomplished the crossing of Bering Strait on ice floes, as is done occasionally today (usually by accident), but large-scale migration would have been virtually impossible had land bridges never existed between Asia and North America. Similar land bridges facilitated the entry of primitive man into the British Isles, across what are now the English Channel and the Irish Sea.

Neither the rise of sea level due to melting of the world's glaciers nor the isostatic elevation of areas once weighted down by ice seem to have had any noticeable effect on more recent history. In Scandinavia, the relatively rapid isostatic uplift and the resulting seaward migration of the shoreline have separated some coastal settlements from the sea in the past few thousand, or even one thousand, years. In areas where there has been local subsidence of the land to augment the eustatic sea-level rise, cities or parts of cities have been submerged, as offshore archeological activities in the Mediterranean have been discovering. In most cases this did not result in abandonment, but merely in retreat to higher ground, a retreat so gradual that the continuity of existence of the community was never interrupted.

The advance and retreat of glaciers on an accelerated but very local scale has had little to do with human history, because very few people have ever lived close enough to the ice front to be affected. Fluctuations of the fronts of mountain glaciers and of the lobes of the remaining ice caps have been directly measured in the past century or more by means of systematic observations, and fluctuations which occurred before direct measurements began can be determined from various kinds of geologic evidence, by radiocarbon dating, and sometimes from archeological and historical evidence. Only in places like Greenland and Iceland and Alaska have settlements ever been seriously affected by ice movement. In southeastern Iceland a few farmsteads near glacier tongues, overridden during temporary ice advances, are now reexposed.

In the Yakutat area of Alaska native traditions concerning ice advances and retreats have been fully corroborated by geologic observations [45]. Geologists have determined that lobes of the Malaspina Glacier once completely filled Icy Bay and Yakutat Bay and then, beginning about A.D. 1400, retreated to positions near or behind their present limits (Fig. 6). A second advance, culminating some time between 1700 and 1791, filled all of Icy Bay and the

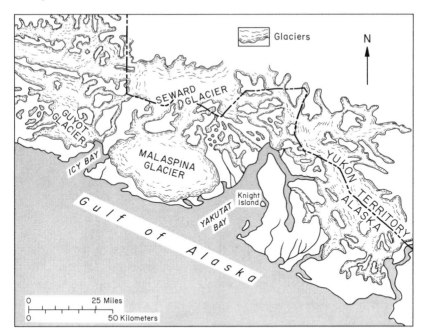

FIGURE 6. *Sketch map of a portion of the coast of southeastern Alaska.
Indian traditions concerning ice advances and retreats in Icy Bay and
Yakutat Bay are in good agreement with the geologic facts.*

head of Yakutat Bay. The ice began to retreat from Yakutat Bay
in 1791, but remained in Icy Bay until about 1904. A native tradition, thought to date from about 1400, tells how Icy Bay and
Yakutat Bay were filled to their mouths while people lived, or at
least hunted, on the ice-free coasts to the east and west. (The name
Yakutat may mean, in Eyak, "a lagoon [or bay] already forming,"
referring to the open water gradually exposed as the ice retreated.)
An archeological site on Knight Island in Yakutat Bay, known as
"Old Town," traditionally was founded before trees had established themselves on the island, which would mean that the island
was settled not very long after the ice had left it exposed. According to another local tradition, a village on the west shore of Icy
Bay, presumably settled after the same retreat, was overwhelmed
by an advance of the ice, presumably the one that culminated in
the eighteenth century.

These traditions are historically accurate insofar as they record
events that are known to have happened. But in trying to explain

why those events occurred, they enter the realm of geomythology. The reason for the retreat that began around 1400 was said to be that Atna Indians, migrating from the Copper River, threw a dead dog into a crevasse as they crossed Yakutat Bay over the ice. The advance that overwhelmed the Icy Bay village was attributed to the fact that some young men jestingly invited the ice to join them in a feast—just as Don Giovanni invited the Commendatore's statue to supper, to his eternal regret. The last retreat from Icy Bay was attributed to the fact that a Tsimshian Indian who died in the area had been buried near the ice front by friends who wished to preserve the body until it could be shipped home—an incident known to have happened some time between 1890 and 1903.

Another slow but perceptible geologic process, sedimentation, has changed the history of certain communities. Many great cities that flourished in earlier days on the shores of the Mediterranean are represented today by mere villages huddled on the ruins of their former grandeur, or have been abandoned completely to archeology. Formerly important ports, they declined as the harbors that were their *raison d'être* became silted up in the natural course of events, or as a result of deforestation which accelerated erosion in the drainage area. The rivers on or near whose mouths they were located had no choice but to dump their sediment when their current was checked upon entering the estuary or bay that constituted the harbor, or when the current slowed as the gradient flattened out on a delta flood plain.

The Phoenicians were the greatest sea traders of the Mediterranean world for many generations, beginning in about 1200 B.C., but their importance dwindled as their harbors, including those of Sidon and Tyre, silted up completely or partially. L. J. Snell [218] has described how sedimentation was at least a contributory cause of the ultimate decline of Troy, Miletus, Ephesus, Priene, Heraclea, Smyrna, and Tarsus. Of these, only Smyrna (modern Izmir), at the head of the Gulf of Izmir, is still a port. It too was on the downgrade because of sediment from the Gediz River (the ancient Hermos), but it has been reprieved, for a few generations at least, by diversion of the river so as to deposit its load near the north edge of the Gulf (Fig. 7).

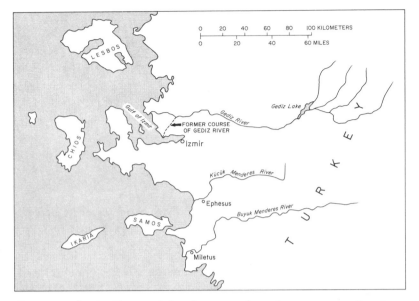

FIGURE 7. *Izmir (Smyrna) has been saved—at least temporarily—from the fate of Miletus, Tarsus, and other once-important cities of the Aegean coast of Turkey, by diversion of the Gediz River. (After Snell, 1963 [218].)*

Tarsus, the city of Paul the Apostle, was once a port connected with the Mediterranean by a short dredged channel. Tarsus still exists and is not exactly unprosperous, but it is now only a farming community, about twelve kilometers inland on a delta plain built by three rivers (Fig. 8). Troy suffered many ups and downs, but the loss of its harbor must have helped bring about its final abandonment; the ruins of Troy are now separated from the coast by four kilometers of sediment deposited by the Scamander River (the modern Kanamenderes). Miletus, with a population that may once have been as large as two hundred thousand, was the principal port and trade center of the Ionian Confederacy. It was situated on a promontory that jutted out into Bafa Gulf (Fig. 9). The Buyuk Menderes, once known as the Meander,* built its delta across the gulf; Miletus' promontory became surrounded by a "sea" of sedi-

* The Meander has lent its name to geologic nomenclature, to designate the bends developed in a stream flowing at grade on its flood plain, which it typifies.

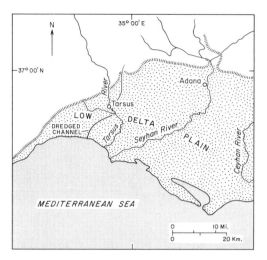

FIGURE 8. *Tarsus, once close to the sea, now lies about seven miles inland, surrounded by delta plain sediment. (After Snell, 1963 [218].)*

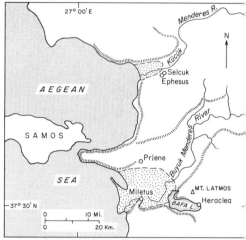

FIGURE 9. *Miletus, Priene, Heraclea, and Ephesus, once port cities, were cut off from the sea by sediment deposited by their respective rivers. (After Snell, 1963 [218].)*

ment, Priene to the north was cut off from the sea, and Heraclea at the head of the gulf was left on a land-locked lake, twenty-eight kilometers from the coast. In the early days of Christianity, Ephesus (see Fig. 9) vied with Smyrna and Pergamum for the honor of being called the first city of Asia. It followed Miletus into oblivion when its estuary became filled with sediment from the Cayster River (known today as Kucuk Menderes) (see Fig. 9). The immediate cause of the abandonment of Ephesus was malaria; as the increasing sediment produced increasingly swampy conditions, the area became progressively less healthful.

Of all the slow changes on the face of the earth, climatic changes—which are geologic changes in a broader sense than those we have been considering—have undoubtedly influenced the development of man more than any other factor. All over the world we can find traces of civilizations that vanished when water became scarce: pueblo ruins, like Mesa Verde, scattered throughout the Southwest, abandoned in the fourteenth century as a result of prolonged drought; or the Harappan civilization of the Rajputana area of India. Iceland's history has been closely tied in with the climate, with two eras of prosperity and independence (870-1262 and 1918—), coinciding with periods of favorable climate, separated by a time when independence was lost to Norway and then to Denmark, coinciding with a period of adverse climatic conditions. Finally, the drastic climatic deterioration in northwestern Europe around 500 B.C. gave the first impetus to the southward movement of Teutonic peoples that eventually led to the fall of Rome. Some scholars maintain that this climatic change was the *fimbulvetr* (the long and awful winter) [91:330] of Norse mythology [241].

It is hard to find geomyths which appear to have been engendered by the slow (in human terms) geologic processes. The Norse *fimbulvetr* may be one example; the Alaskan traditions mentioned above, another. The Alaskan events are comparatively recent. In fact, there may still be people living who remember when Icy Bay was last filled with ice. There, as in the Mediterranean cities which declined in importance as their harbors disappeared, or in the Scandinavian settlements left inland as a result of isostatic uplift, the inhabitants had plenty of time to adjust to the changing conditions and move elsewhere, if necessary. Even if there were no written records, there has hardly been time for the Alaskan ice fluctuations to have been forgotten.

Most of the great Mediterranean cities of olden times, now abandoned, are remembered chiefly through historical records. Had they existed before written history, how many of them might not have been totally forgotten? Troy, to be sure, became legend, but that was Homer's Troy, the sixth of the nine cities built on the site. Greco-Roman Troy, the one whose end can be attributed at least in part to sedimentation, lasted until the fifth century A.D. The Troy that was remembered, however, was the one whose violent end at the hands of the Greeks was commemorated in Homer's poetry, and it was long believed to have been fictional. The younger Troys, which suffered a slower death, whether due to man or nature or both, were completely forgotten for centuries. Likewise, the cryptic inscription referring to the demise of Old Mataram in Java, discussed in the previous chapter, gives no hint of the slow decline which had already set in as a result of the silting up of Bergota's harbor, but only of the violent final disaster.

Thus the very paucity of examples serves to emphasize that the influence of gradual but perceptible geologic processes on legend is minuscule, and that only the spectacular events beget euhemeristic legends. Man adapts to the slow-but-sure changes so naturally that in most cases he forgets them completely.

4 /

Landform Lore

There were giants in the earth
in those days.

GENESIS 6:4

THIS CHAPTER ASSEMBLES SOME TYPICAL examples of folklore associated with landscape features. Nearly always, these features were created by forces whose operation predated man's appearance in the region, and the geomyths therefore are purely etiological. Actually, the title "Landform Lore" is not as comprehensive as it might be, for we include here folklore associated not only with large-scale landscape features, but also with mineral deposits and even a few individual minerals.

In seeking an explanation for topographic features too large to have been created by ordinary men, it was natural for the primitive imagination to attribute them to the work of a race of beings of superhuman size and strength. Thus all over the world we find legends which attribute mountains, hills, large boulders, lakes, islands, and almost any landform one can name to the efforts of giants; or in some cases these landforms are regarded as the giants themselves, turned to stone. Popular belief in giants was often reinforced or perhaps even created by the finding of huge bones of mastodons, mammoths, or other large extinct animals in glacial deposits.

Germanic tradition is particularly rich in giant lore [91, 92]. According to one German myth the giants marred the pristine

37

smoothness of the newly-created earth by clumping about on its still-soft surface; the giantesses wept to see the gouges made by their husbands' clumsiness (the river valleys), and their tears formed the streams [92:231]. This is as good an example as any to illustrate how far from fact etiological folklore usually is. Geologists are still only speculating as to when and how the original crust of the earth was formed, but I think all would agree that it was not originally smooth and soft. Then once the atmosphere had formed (by accumulation of gases given off from the interior, as is still going on through volcanoes), the processes of erosion and deposition could begin. Streams do not merely occupy their valleys, they create them. The first rains falling on the earth must have drained away via low spots in the primitive crust, establishing the first stream valleys there. Throughout geologic history, river valleys have been born, have grown through youth, maturity, and old age, and have either been rejuvenated or "drowned" as the land has risen or sunk.

The Germanic race of giants could move about only in darkness and fog; if touched by the sun's rays they turned to stone. The Riesengebirge ("Giant Mountains"), one of the Sudeten ranges on the border between Prussian Silesia and Czechoslovakia, were said to be giants who failed to take cover in time and were petrified by sunlight [91:232]. To the geologist, the Riesengebirge are block mountains—mountains carved by erosion from large uplifted earth-blocks bounded on one or both sides by fault-line scarps—composed of folded gneisses and granites. They are the highest and most rugged of the so-called Mittelgebirge ("Central Mountains"), so their name and the legend may have been inspired merely by their size.

The Siebengebirge are seven hills on the right bank of the Rhine near Bonn. Seven giants are said to have been hired to dig a canal (the Rhine), and upon completing the work knocked the dirt off their shovels to form these seven piles. In reality the Siebengebirge are volcanic in origin, a series of trachyte domes and basalt plugs formed in Tertiary time. A volcanic dome is a steep-sided protrusion of viscous (that is, thickly liquid rather than runny) lava which bulges up over and around a volcanic vent. A volcanic plug is the solidified filling of the vent of an extinct volcano. A dome is pushed above the surrounding terrain and remains higher if its material is more resistant to erosion than the material it intrudes, which is often the case; a plug, being more resistant usually than the rest of the material constituting the volcanic mass, will stand out as an isolated column or crag after the volcano has been eroded away.

A North Frisian tradition attributes the whiteness of the White Cliffs of Dover (Plate 6)—which are white because they are made

PLATE 6. *The White Cliffs of Dover. (Photo by Donald E. Hattin, April 1969.)*

of chalk—to a giant's ship which nearly became stuck when it tried to pass through the English Channel from the North Sea. The crew soaped the sides heavily, especially the starboard side where the rocks rose in sheer cliffs, and the ship just managed to squeeze through; but so much soap was scraped off on the cliffs that they remained white forever after, and the waves dashing against them are usually foamy [91:235].

Torghatten is a hat-shaped promontory on Torget Island (Plate 7) off the west coast of Norway, about 150 miles north of Trond-

PLATE 7. *Torghatten, off the coast of Norway north of Trondheim. (Photo by Knut Aune Kunstforlag A/S, Trondheim. Courtesy of the Geological Institute, Technical University of Norway, Trondheim.)*

heim. About 400 feet above its base it is penetrated by a natural tunnel 550 feet long, up to 250 feet high and 56 feet wide. It would be surprising if such a unique landform did not have a legend attached to it [91:233]. Thus, a giant named Senjemand fell in love with the beautiful giantess Juterna-jesta, who lived about eighty miles away, but she refused him scornfully. In a rage he shot one of his giant stone arrows at her, but her lover, Torge, flung his hat and intercepted it. Senjemand fled on horseback, but the rising sun caught him, and he and his steed, together with Torge's hat, were turned to stone. Torghatten is that hat, the tunnel in the mountain is the hole in it made by the arrow, a natural obelisk in the vicinity is the arrow, and the island called Hestmona, on the Arctic circle, is the petrified horseman.

The real reason for the hole in Torghatten is erosion along a joint. Joints are fractures or partings which abruptly interrupt the physical continuity of a rock mass. They are formed as a result of stress, and constitute lines of weakness along which weathering and erosion can proceed faster than in the adjacent solid rock. The granitic rock of which Torghatten is composed is traversed by several systems of joints. At a time when the land stood much lower relative to the sea than at present (during an interglacial period, and before isostatic uplift had occurred), waves pounded against the island, incessantly hurling a battery of stones and sand against it. All around the island a notch was cut, marking the base of the "crown" of the hat, while ever-deepening sea caves developed at both ends of a weak joint zone and finally met to form the tunnel [199].

An equine giant is credited with the formation of Ásbyrgi, a beautiful and unusual horseshoe-shaped depression in northeastern Iceland (Plate 8). The Ásbyrgi depression is supposed to be the impression of one of the hooves of Sleipnir, the eight-legged steed ridden by Odin, the Scandinavian counterpart of Zeus. The depression is really a fossil riverbed and waterfall, eroded by the river Jökulsá-a-Fjöllum, which flowed there (in late Glacial and early Postglacial times) until it was naturally diverted to its present course. Since the diversion took place after the glaciers had withdrawn from the area, the change in course cannot be attributed to disruption of the drainage by ice tongues [243:164]. The change was probably caused by a meltwater flood due to subglacial volcanic activity under the northern part of the Vatnajökull ice cap (see Fig. 3).

Devil's Tower, or Mateo Tepee as the Indians called it, is a prominent landform jutting above the surrounding countryside in northeastern Wyoming. Naturally the Indians had their explanations of its origin [154]. According to the Kiowas, seven little girls who were playing some distance from their village were pursued by bears. Realizing that they could not reach the safety of the village in time, they jumped onto a low rock and prayed to the rock to save them. Immediately the rock began to shoot upward, and when its top reached the sky the children were turned into the seven stars which we know as the Pleiades. The distinctive vertical grooves down the flanks of the Tower are the marks left by the bears as they clawed at the rock in a vain attempt to reach their intended victims.

The Cheyenne legend is somewhat different, but it too attributes the peculiar features of the Tower to the clawing of a bear. According to the Cheyennes, the wife of the eldest of seven brothers was carried off by a giant bear. They all went to the res-

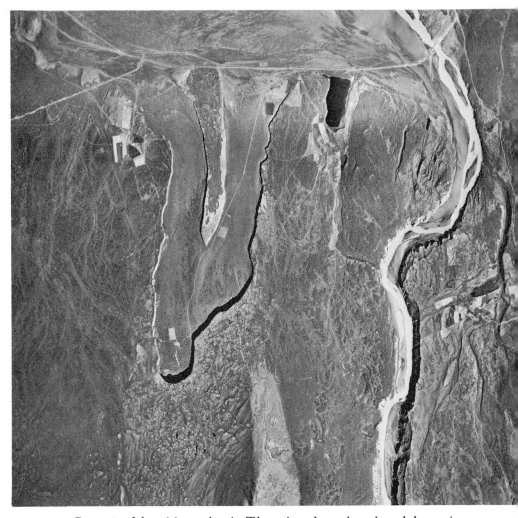

PLATE 8. *Ásbyrgi from the air. The unique horseshoe-shaped depression in northern Iceland is the gorge below a "fossil" waterfall, cut when the river Jökulsá a Fjöllum (right) flowed to the west of its present course. From an aerial photograph. (Reproduced by permission of the Iceland Survey Department, Reykjavík.)*

cue, and with the help of the youngest brother, who was a powerful medicine man, succeeded in freeing her while the bear slept. The bear awoke, and summoning his followers, gave chase. At the place where Devil's Tower now stands, the youngest brother sang a magic song, and a small rock which he carried grew into the Tower, taking the party safely out of the bears' reach. The brothers killed all but the leader of the bears, whose attempts to reach the band on top (depicted in Plate 9) resulted in the characteristic fluting of the Tower. Eventually the youngest brother succeeded in killing the giant bear, and summoned eagles which carried the group to the ground.

The geometric perfection of the sheaf of columns constituting Devil's Tower suggests an artificial origin, but those columns are entirely the work of nature nonetheless. Columnar jointing, of which it is a classic example, typically forms in some kinds of lava as it cools, shrinks, and cracks. The cracks begin to form on the cooling surface, in exactly the same way that mud cracks form upon drying and shrinkage of the surface of a mud flat. Shrinkage cracks tend to mark off six-sided areas because the hexagon is the shape which encompasses more area within a given perimeter than any other when the shapes are completely contiguous (otherwise, the circle contains the greatest area within a given circumference). As a lava lake cools, the cracks extend downward from the surface to form long, close-packed columns whose regular shape may be revealed subsequently by erosion. Not all such columns are formed in lava lakes however. They can develop as easily in sills intruded between strata, like the well known Palisades across the Hudson River from New York City, in dikes cutting across the "grain" of rocks, or in certain kinds of volcanic rocks called "ash-flow tuffs." In all cases the columns are perpendicular to the cooling surface, whether it be horizontal or vertical or at an angle. Devil's Tower, made of the rock called phonolite (because it "rings" clearly when struck), is either the neck of an old volcano now eroded away, or a small plug of volcanic material intruded into softer rocks now eroded away. An even more famous example of columnar jointing is the Giant's Causeway, a promontory on the north coast of County Antrim in Northern Ireland. The basalt columns there are fifteen to twenty inches across (Plate 10) and sometimes as much as twenty feet in height. Legend has it that it once extended all the

PLATE 9. *The legend of Devil's Tower, Wyoming. (Courtesy of the U.S. National Park Service.)*

PLATE 10. *Tops of basalt columns, Glants' Causeway, County Antrim, Ireland. (Photo by Donald E. Hattin, January 1969.)*

way to Scotland, providing a route over which giants traveled in the days of old.

The most spectacular mountains on wild and picturesque Skye, largest of the Inner Hebrides of Scotland, are the Cuillins, which rise precipitously from sea level to 3,309 feet at their highest point. Visible in the same panorama from some points are the more subdued Red Hills, more rounded in contour and distinctly reddish brown in color. Although the Cuillins are higher, snow rarely clothes their somber slopes for more than a few hours, but on the Red Hills it may linger for days. The Cuillins were named for a giant, but the story of how that came about is more complicated than the simple cause-and-effect kind of legends like those above [231].

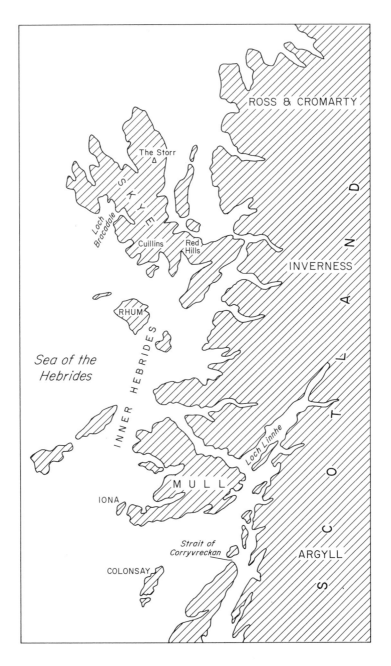

FIGURE 10. *Sketch map of the Inner Hebrides, showing places mentioned in the legend of the Cuillins.*

Cailleach Bhur, or Winter (literally, the Hag of the Ridges), is credited with having created Scotland by dropping peat and rock into the sea. Originally the area between the Red Hills and Loch Bracadale on Skye (Fig. 10) was a great plain, and there Cailleach Bhur used to come while her washing boiled in the whirlpool of Corryvreckan; to dry the clothes she spread them on Storr, a 2,360-foot peak in the northern part of Skye. The Hag held prisoner a maiden loved by Spring, who called upon the Sun to help free her. The Sun flung his fiery spear at the Hag as she walked the moor one day, but missed. Where the spear hit the ground a huge blister grew and swelled until it burst and discharged a molten, glowing mass, which so frightened Cailleach Bhur that she ran away and hid forever. The glowing mass congealed to form mountains which the snows have never been able to conquer. Later the mountains were named for the Irish giant Cuchullin, to commemorate his battle with a local warrior goddess, in which neither prevailed.

The Cuillins (Plate 11) are made of gabbro, a type of rock which crystallizes from molten matter, but far below the surface rather than on it. In fact, they are the deep roots of volcanoes which were active in Tertiary time. The Red Hills (Plate 12) are made of granite, another deep-seated igneous rock. Being of different min-

PLATE 11. *Scurr nan Gillean, in the Cuilins of Skye, from the Sligachan track. (Photo by R. Thompson, September 1962.)*

PLATE 12. *The Red Hills on the island of Skye. (Photo by C. J. Vitaliano, September 1969.)*

eralogic composition, granite and gabbro weather differently. The snow lingers longer on the Red Hills, even though they are lower in elevation, merely because their slopes are less steep.

The notion of a blister forming the earth's surface and discharging a red-hot mass could only have been imagined by some early inhabitant of Skye, for volcanism in the Hebrides has been extinct for millions of years; nevertheless it could almost be an eye-witness account of the formation of a volcanic dome. The Tristan da Cunha eruption involved the formation of a lava dome, and another was seen to form rather recently in Japan. On December 28, 1943, a series of sharp earthquakes began to shake the shores of beautiful Lake Toya on Hokkaido, Japan's northernmost island, and a huge blister began to swell the ground near the volcano Usu. Within six months it had risen to 165 feet above the normal ground level, and on June 23, 1944, it burst and erupted volcanic ash. It continued to grow, and ash continued to be erupted in small explosions. Finally, in November 1944, a dome of red-hot but solid lava started to protrude through the still-rising bulge. When the dome, named Showa-Shinzan ("the-new-mountain-of-the-reign-of-Showa"), reached a height of 990 feet above the original level of the land, it stopped growing; that was in September 1945, and today Showa-Shinzan is still vigorously discharging vapor (Plate 13). If this had happened

PLATE 13. *Mount Showa-Shinzan. Hokkaido, Japan, a new volcanic dome created in an eruption in 1943-44 in much the same manner in which legend (erroneously) explains the formation of the Cuillins of Skye. (Photo by C. J. Vitaliano, September 1970.)*

long ago, and was described only in some local legend of the Ainu people who inhabited Hokkaido before the Japanese, surely we would think it very much exaggerated, or even fictitious.

Classical mythology offers two versions of how the Atlas Mountains came to be, both involving giants. One of the Titans who sided with Cronos in his war with Zeus was Atlas, a nephew of Cronos. When the victorious Zeus punished his adversaries, Atlas was condemned to stand for eternity supporting the heavens on his shoulders. The other version is part of the story of Perseus. After slaying the Gorgon, Perseus, still wearing Hermes' winged sandals and carrying the Gorgon's head in a bag, flew far and wide until toward nightfall he found himself near the western limit of the earth. There he sought shelter for the night with King Atlas, a man of huge stature and very rich to boot, but was turned away because

Atlas feared he would rob him of his prized golden apples. At this unheard-of breach of the conventions of hospitality, Perseus averted his gaze and displayed the Gorgon's head, and Atlas was straightway petrified. "His beard and hair became forests, his arms and shoulders cliffs, his head a summit, and his bones rocks. Each part increased in bulk till he became a mountain, and (such was the pleasure of the gods) heaven with all its stars rests upon his shoulders": thusly Bulfinch [24:98] describes the transformation. Both versions reflect the geographic fact that the Atlas Mountains, at the western limit of the world known to the ancient Greeks, appeared to hold up the sky.

Not all mountains are explained as the work of giants or as transformed giants; some are personified, as in this tale from our Pacific Northwest [36:13]: The peaks of the Cascades Range were once people. Pahto, whom we now call Mount Adams, and Wyeast, whom we know as Mount Hood, fought over a girl. Pahto lived north of the Columbia River and Wyeast south of it (Fig. 11), but at the time there was a bridge over the river and they frequently crossed it to fight, now on one side and now on the other. Old Coyote summoned the other peaks to help stop the quarreling. They all began to march northward to a big council meeting, but before they could get there Coyote caused the bridge to fall in a last effort to keep the two antagonists apart. When the mountain people heard that the bridge was down, they stopped in their tracks, and there they stand today—Mount Jefferson, the Three Sisters, and all the others. Black Butte had sat down along the way to rest, while her husband, Green Ridge, lay full length beside her; the sun was hot, and sweat poured off Black Butte in two streams which met to form the Metolius River. The Metolius rises near the foot of Black Butte, an extinct volcano 6,436 feet high, and flows roughly northward in the initial part of its course. Green Ridge, just to the northeast of Black Butte, is a long north-south trending ridge parallel to the Metolius, made of Tertiary basalt flows of the Columbia River series.

The bridge over the Columbia River mentioned in this geomyth appears at first glance to be a memory of some real geologic event. Although the individual legends vary in detail, all the tribes living along the Columbia River agree that it once flowed through

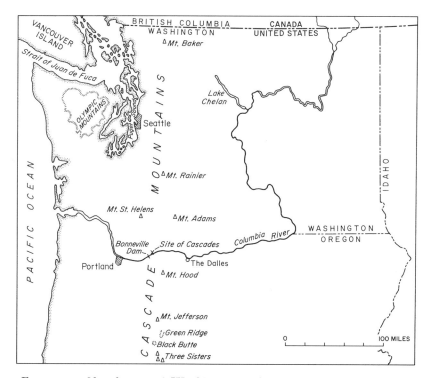

FIGURE 11. *Sketch map of Washington and northern Oregon, showing the location of places mentioned in various legends concerning mountains, lakes, and other geographic features of the area.*

a tunnel or under an arch known as the Bridge of the Gods [36:20-24], which either crashed into the river during an earthquake caused by the struggle between Mount Hood and Mount Adams, or was demolished by rocks hurled by these volcanoes at each other, some of which also fell into the river farther east to form the Dalles (see Fig. 11). A natural bridge spanning the river at this point is a structural impossibility. However, an enormous rock slide, from mountains of which Table Mountain and Red Bluffs are the remnants, did once completely block the Columbia here. Upstream from this barrier trees could be seen under twenty to thirty feet of water, their branches sometimes extending close enough to the surface to pose a threat to canoes and other craft. The Cascades of the Columbia marked where the river was cutting its way through the slide material. Today a modern steel structure, also called the

Bridge of the Gods, crosses the river at the site; the rapids, however, have been submerged beneath the waters backed up by Bonneville Dam. The Dalles are flat-topped, vertically-walled islands. River channels of Dalles type are peculiar to the Columbia River basalt plateau; they are formed by rivers of large volume with high gradient flowing over closely and vertically-jointed rocks, which erode by plucking rather than abrading [17] and thus leave islands unusual enough to provoke speculation as to their origin.

Whether the Indians actually witnessed the landslide which presumably formed the Cascades is uncertain. On the basis of Indian tradition the Bridge of the Gods is supposed to have collapsed "in the time of our grandfathers"; taking "grandfathers" literally, it was estimated that the event happened some time between 1750 and 1760.* But from the geologic evidence, the landslide could have happened as much as a thousand years ago, in which case the expression "our grandfathers" should be taken only figuratively, to mean "our forefathers." Had the event occurred as recently as the middle of the eighteenth century, I feel certain the tradition would probably reflect the geologic facts somewhat more closely than does a mythical bridge. As it is, except for the implication that the Indians witnessed some activity of Mount Hood and Mount Adams, the Bridge of the Gods, like the explanation of the Dalles, seems to be a purely etiological invention.

American Indian lore is rich not only in myths accounting for mountains, but also in those explaining the presence of other kinds of landforms, water bodies, rivers, and various other features of

---

* The difficulty of dating geologic events on the basis of tradition, even when the number of generations involved is known, is illustrated by attempts to date the last eruption of Haleakala on Maui in the Hawaiian Islands, lava flows from which form Cape Kinau. The first calculation of the time of the eruption, based on reports by grandchildren of eyewitnesses, allowed thirty-three years per generation and yielded a date of 1757; a recalculation, taking twenty-five years as the more probable average length of a Hawaiian generation, put the date at 1770. By comparing early anchorage charts of the coast in this region, B. L. Ootsdam [172] recently demonstrated that the latter was still in error by about twenty years. The chart drawn in 1786 by the French explorer La Pérouse (for whom the bay south of Cape Kinau is named) showed a shallow embayment uninterrupted by the prominent bulge of Cape Kinau; that drawn by the English navigator Vancouver in 1793 definitely shows the cape, which therefore must have been formed in 1790, give or take three to four years.

the landscape. In Washington state they explain Puget Sound and the Cascades Range in this way [36:25-26]: When the world was very young, the land was flat where the Cascades now rise. Rain was not yet known, but the moisture needed for trees and plants came up out of the ground. Then for some reason it stopped coming in the region which is now eastern Washington. A delegation was sent to Ocean in the west, to beg him to send water. Ocean responded by sending his children, Clouds and Rain, and soon the land was fruitful again. But the people were greedy and would not let Clouds and Rain go home, and kept them busy filling pits they dug to hold more and more water. Ocean sent word that the people could count on him for water whenever they needed it, but still they would not let Clouds and Rain go. So Ocean prayed to the Great Spirit to punish the people. The Great Spirit leaned down from heaven and scooped up a great mass of earth, and with it fashioned the Cascades; Ocean flowed into the hole whence the material had been removed, making Puget Sound. The land east of the Cascades dried up, for Ocean sends little moisture over the mountains; all the water left to the people on the east side of the range is that in the pits their forefathers dug, the largest of which is Lake Chelan (see Fig. 11).

In reality, the Cascades were built by volcanism, the activity lasting throughout Tertiary and Quaternary times and still continuing—six of the peaks are technically classed as active [39]: Mount Baker, Mount Rainier, Mount Saint Helens, Mount Shasta, Cinder Cone, and Lassen Peak. (An active volcano is one which is known to have erupted in historical time. The recorded eruptions of all except Lassen Peak have been very feeble indeed, hardly worthy of the name.) Glaciation of the high peaks during the Ice Age helped shape the sculptural details of this imposing range, and Mount Rainier still retains the largest system of mountain glaciers in the United States outside of Alaska. Puget Sound is a large estuary,* and Lake Chelan occupies a river valley which was overdeepened by one of the valley glaciers during the Pleistocene period. But the Indians seem to have been conscious of the real reason why the land east of the range is dry, for the Cascades pre-

---

* Estuary: the widened channel at the mouth of a river, in which there is a marked tidal action; usually formed by submergence ("drowning") of a river valley—or in this case, valleys.

sent a barrier over which moisture-laden winds have to rise, with the result that they become chilled in the higher atmosphere and precipitate the greater part of their moisture on the west slopes.

A very different myth [36:70-71] connected with the origin of Lake Chelan starts out similarly: Once there were no mountains or lakes in this part of the country, only a grassy prairie with abundant game. Along came a monster which ate or frightened away so many animals that the people starved. The Great Spirit heard their prayers and killed the monster, but twice it came back to life. After killing the monster a third time the Great Spirit struck the earth with his great stone knife; the earth shook and a huge cloud descended and hid the land. When it cleared, everything was changed. Where there had been a plain there was now a range of lofty mountains; deep canyons marked the places where the rocks and dirt had been removed to make the mountains. The Great Spirit threw the monster's body into the longest and deepest of the canyons and filled it with water to form Lake Chelan. This time the monster stayed dead, all but its tail, which still thrashes around and makes such big waves that the Indians avoid paddling their canoes on Lake Chelan. (Lake Chelan is fifty-five miles long but not more than a mile and a half wide, and it is easy to understand how it might be treacherous to canoes when the wind blows strongly in certain directions.)

Hell's Canyon of the Snake River, deeply entrenched in a high plateau, is one of the most inaccessible gorges in the United States, deeper in places than the Grand Canyon of the Colorado. On its Idaho side rise the Seven Devils Mountains, and farther to the west are the Blue Mountains (Fig. 12). This Nez Percé Indian myth [36:47-48] purports to explain some of the geographic features of this area: Long ago the Blue Mountains were inhabited by seven giant brothers who terrorized the ancient people of the region. Every year the brothers traveled eastward seeking children to eat. Coyote called all the digging animals together and had them dig seven very deep holes across the giants' usual path to the east, and filled the holes with boiling reddish-yellow liquid. When the time came for their annual foray, the giants marched along with their heads in the air, confident in their superior size and strength. Into the seven holes they stumbled, and much as they struggled, splat-

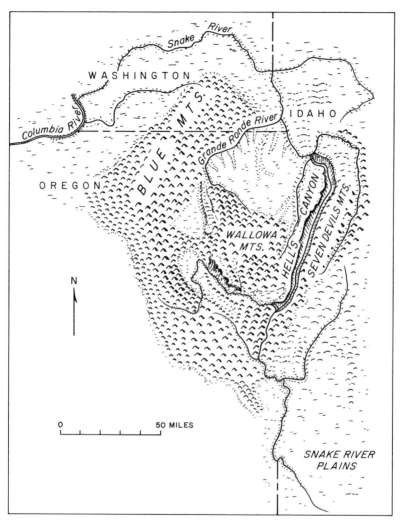

FIGURE 12. *Geography of the Hell's Canyon–Blue Mountains region.*
(After D. C. Livingstone, 1928.)

tering the liquid a day's journey in every direction, they could not
extricate themselves. Then Coyote changed them into seven mountains, standing high to remind people that retribution is sure to
follow wrongdoing. He struck the earth and caused a deep gash to
open at the foot of the new peaks, to prevent any others of the
giants' family from venturing across from the Blue Mountains. The
liquid which was splattered during the giants' struggles became
the copper which is mined in the area.

55

Geologically, the Seven Devils Mountains, rising to altitudes of more than nine thousand feet above sea level and standing well above the level of the surrounding plateau, were carved by erosion from an uplifted fault block. The Blue Mountains are also elevated blocks of basalt, rising in a series of steps, with a granite core protruding through the basalt in the highest part. The drainage history of the area is complicated. The Snake River seems to have begun as a tributary to the Grande Ronde River. It eroded its way headward along lines of weakness, following northeast-trending faults and the general structural trend of the older rocks in the area, and as it cut its way into the slowly rising plateau to form Hell's Canyon it captured an increasingly great share of the drainage from the Wallowa Mountains and from the mountains on the Idaho side, until eventually it short-circuited much of the drainage of the Grande Ronde [142].

In addition to the various authentic Indian myths and legends associated with landforms, North America has a modern brand of geographic folklore (or, as it has been more accurately termed by R. M. Dorson, "fakelore") in the tales concerning the giant woodsman Paul Bunyan [59], hero of the logging camps, and his companion, Babe, the Blue Ox. Numbered among the manifold exploits of this team, for instance, is the digging of the St. Lawrence River, because without a boundary, people couldn't be sure whether they were in the United States or Canada. Paul and Babe accomplished the mighty task in just three weeks, using a scoop shovel as big as a house. The dirt they excavated was dumped on Vermont, where it formed the Green Mountains. When the million dollars he was promised for the job was not forthcoming, Paul threatened to fill in the new ditch again, and threw in a few shovelfuls of dirt just to prove he meant it; the money was promptly paid, but those few shovelfuls remain as the Thousand Islands.

Once when Paul and Babe were in the northwest, Babe was frightened by the roar of Spokane Falls and ran away with the provision sled, dragging the swamp hook. The hook gouged out the Columbia River Gorge and finally caught fast in the Cascade Mountains. When Paul pulled the hook free, water poured into the bottom of the hole it left. He started to throw in some rocks to stop the leak, but had to stop because the blue ox was getting too

PLATE 14. *Crater Lake, Oregon, showing Wizard Island, a young volcanic cone within the caldera depression. (Photo by the author, August 1959.)*

nervous. The hole filled up, forming Crater Lake. One of the rocks dropped by Paul is Wizard Island in that lake (Plate 14).

In order to deliver the logs he and his crew felled in northern Minnesota, Paul dug out the Mississippi River to the Gulf of Mexico. The dirt that flew over his right shoulder made the Rocky Mountains, and that tossed over his left shoulder made the Appalachians. When the task was finished, he flung aside the shovel, which became the Florida peninsula, and a mitten he dropped on the way back to his northern camp became the lower peninsula of Michigan, thumb and all.

Paul and Babe were also responsible for the Grand Canyon. Before they came along, the Colorado River was known as Old Contrary, because in some stretches it was a mile wide and a foot deep and in others a mile deep and a foot wide. With Babe hitched to a bulltongue plow, he widened it where it was deep and deepened it where it was wide, just to even things up.

Almost any major or minor topographical feature of North America has been or could be woven into the Paul Bunyan stories. Resemblances to other legends in which giants create curious landforms are obvious, but there is one essential difference between Paul Bunyan and the giants of earlier cultures: the exploits of Paul and

Babe and the rest of their crew are a purely literary creation, and none but the most innocent hearts has ever believed they existed. These stories are in a class with tall tales and Santa Claus.

Possibly in the same vein as the Paul Bunyan stories, but perhaps real folklore, is a cynical bit from Montenegro, the least developed part of Yugoslavia. On the whole Montenegro is difficult mountain country, much of it in barren karst.* No wonder, therefore, that they say that when God finished creating heaven and earth, he dumped the unusable scraps in a heap, and that made Montenegro.

There are some mountain myths which have been inspired by a resemblance to familiar objects. Such is the Takitimu Range in southern New Zealand, which the Maoris considered to be the overturned hull of one of the original fleet of canoes that brought them from their legendary homeland of "Hawaiki,"† turned to stone; its toothed crest represents the broken keel [40:153]. The Takitimu Range is made up mainly of layers of various kinds of volcanic rocks that were once erupted onto the sea floor in a geosyncline (a deep subsiding trough), together with some interbedded sedimentary rocks of the type called graywacke, derived from such volcanic rocks; the whole series was later folded and uplifted above sea level, and because of differences in the resistance of the different kinds of rock involved, erosion and weathering have produced the present shape which inspired the folklore. Indonesia also has an overturned boat, the volcano Tangkuban Prahu overlooking Bandung, whose smooth outline is shaped like the bottom of one of the native proas. Its story is told in the chapter on volcano legends.

Other kinds of landform legends from times and places remote from each other illustrate the wide scope of the human imagination.

* The geologic term *karst* (the German name for the Carso plateau in northern Yugoslavia, which is the type locality for this kind of topography) denotes a limestone area where the landforms have been determined predominantly by solution and underground drainage, limestone being the most soluble of all rocks. The collapse of underground caves or channels results in a surface pock-marked by sinkholes; streams disappear underground and reappear elsewhere, and the relief is generally chaotic.

† Believed to have been in the Society Islands, not Hawaii.

A peculiarly shaped block of rhyolite, hollowed out on one side as though artificially, lies at the foot of Ngatuku Hill on the Rotorua-Taupo road in New Zealand, where it has rolled down from its former outcrop on the side of the hill. A man named Hatupatu is said to have taken refuge in this rock when pursued by the witch Kura-of-the-Claws [40:119]. When he pronounced the words "Matiti, matata," the Maori equivalent of "Open sesame," the rock opened to admit him and then closed. But Kura waited, and when he emerged the chase continued, ending when Hatupatu successfully leaped across a wide pool of boiling mud which Kura did not see until too late. The unnatural-looking hollow in Hatupatu's Rock is just an erosional feature often found in the face of such blocks. Grooves in a boulder in the same general vicinity, said to be the marks made by Kura's talons when she clawed at Hatupatu during the flight (but he ducked behind the boulder just in time), are also presumably erosional, due to etching out of weaker parts of the rock as it weathered.

The well-known Maori myth of the origin of New Zealand [198:136] illustrates how a folk tale may contain some elements of geologic truth purely by coincidence. While fishing one day with his magic hook the demigod Maui (who also figures in many Hawaiian tales) caught the doorway of the house of Tonganui, son (or in some versions, grandson) of the sea god. Tugging powerfully at the line, Maui pulled up not only the house but also the smooth and shiny land beneath it. Maui beached his canoe and went ashore to make peace with Tonganui, cautioning his brothers to remain quietly behind. No sooner had he disappeared from their sight, than they disobeyed and ran about, hacking at the new land with their knives and claiming pieces of it for themselves. Now this "land" was really the back of a giant fish which had been peacefully sleeping. When attacked it thrashed about, and the smooth back was broken into rugged mountains and valleys and rough rocky coasts. Maui's fish hook (Te Mahia a Maui) is the point on Hawke Bay now known as the Mahia Peninsula (see Fig. 2). It would be tempting to believe that the story of Maui reflects an awareness of the active mountain building processes that have uplifted New Zealand from the sea in geologically recent times. To the trained eye there is abundant physiographic evidence of recent uplift, particularly the magnificent series of raised terraces developed along the rivers of the South Island which flow into the Pacific Ocean. However, the Maoris reached New Zealand only some eight hundred years ago, which is hardly enough time for the slow process of uplift to have been perceptible. In any case, they brought Maui with them when they emigrated from "Hawaiki," for many Polynesian peoples have legends about him, including his having fished islands up from the depths of the sea.

The aborigines of Australia have a number of legends concerning the origin of rivers and lakes, which is not surprising in view of the fact that water is a scarce and highly prized commodity in much of that continent. The Murray River is the chief river, rising in the Australian Alps and flowing for about twelve hundred miles into Encounter Bay through Lake Alexandrina, near Adelaide. According to legend, an earthquake created a narrow rent in the earth in which a tiny stream flowed when it rained. Then, in another earthquake, an enormous fish forced its way to the surface from somewhere deep in the earth. Finding itself much too large for the little stream, it stuck its head into the ground and bulldozed its way toward the sea, widening its path as it went along with mighty strokes of its tail. Water flowed from the depths whence the fish had emerged, filling the valley behind it to form the Murray. At Lake Alexandrina the Ruler of the Heavens seized the great fish, cut it into little pieces, and threw the pieces into the river, where they became the various kinds of fish which live in its waters today [195:73].

The Murray River (which played an important part in the early development of southeastern Australia, where it and its tributaries, navigable by paddle steamer, were the main arteries of travel and commerce) flows in rugged gorges in its upper reaches, then through rolling foothills, and emerges onto a flat, featureless plain —or rather, plateau—underlain by sediments deposited in an inland sea in Tertiary time. As the area rose very slowly from the sea, the lower course of the river became entrenched* in the plateau. From the ground one cannot realize that the valley is there until one ap-

---

* An entrenched stream is a narrow meandering trench cut in a wide-open, flat-bottomed trough which is sunk well beneath the general surface of the adjacent upland.

proaches the edge of the cliff. The valley walls are almost vertical and between one hundred and two hundred feet high; the valley floor is flat, with the stream winding between low river terraces. The whole valley gives the impression that the earth has been wrenched apart, and that if the walls were pushed together the crack would close up. Nothing of the sort has happened, however; the stream itself has carved out the trench bit by bit. But it is very easy to understand how the aborigines imagined a monster bulldozing the lower Murray valley, particularly when one sees it from the air. Lake Alexandrina is a typical coastal lagoon* at the river's mouth. The long sand bars separating it from the sea effectively bar navigational access to the sea.

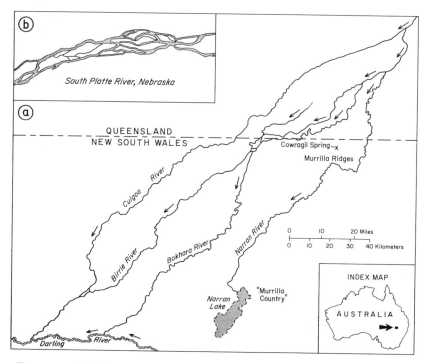

FIGURE 13. *(a) Abnormal drainage pattern in the Narran Lake area, Australia. The entire drainage system anastamoses like the channels in an individual braided stream (b).*

* A lagoon is a salt-water or brackish lake separated from the open sea by marine deposits such as sand bars. The name is also applied to the expanse of water behind the coral barrier reef of a tropical or semitropical island, or in the center of an atoll.

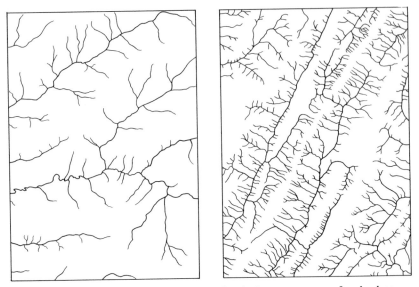

FIGURE 14. *Dendritic (a) and trellis (b) drainage patterns. In the latter, the main streams have developed along a series of parallel outcrops of softer rocks while their tributaries flow at right angles to them, down the slopes of the ridges of harder rock left standing between the main valleys. The geologic structure controls the drainage pattern just as a horticulturalist controls the growth of an espaliered tree; without such control, the drainage pattern resembles the normal dendritic pattern of tree growth, as shown in (a). (From an illustration in* Principles of Geomorphology, *by W. D. Thornbury [244]; reproduced with the permission of John Wiley and Sons.)*

Narran Lake in Australia, near the Queensland border in New South Wales, is accounted for in this fashion [178:27]: As the two young wives of Baiame, the Great Spirit, were bathing at Cowragil Spring, crocodiles grabbed them and carried them off along an underground watercourse leading to the Narran River, the water draining away after them. Baiame pursued, taking a shortcut across a bend in the river—where his tracks are marked by the *murrillas*, pebbly ridges stretching down to the river. At the end of the river he caught up with the crocodiles and fought and killed them. Lashing about in their death throes, they created a great hollow, which the water quickly filled, and forever after the Narran has overflowed into that depression at flood time.

The drainage pattern in this part of Australia is decidedly abnormal (Fig. 13). Normally, small streams join larger to form tree-like patterns (Fig. 14a). Even when there is strong structural control, allowing erosion to progress more rapidly along the outcrop of weaker beds or along joints, the result is still tree-like, though in that case the pattern is analogous to that of an espaliered tree, forced to bend to a shape it would not naturally assume if left free (Fig. 14b). The only time a stream splits up into separate branches is when it is choked by its own heavy deposits, as on a delta (see Fig. 38) or outwash plain (see Fig. 3) or alluvial fan,* or between its own banks at times of low water. Because the individual channels, particularly in the last case, divide and rejoin like braided strands, this type of pattern is called a braided stream (see Fig. 13b).

In the Narran Lake area, as in many parts of the Australian interior, the topography is exceptionally flat. As a consequence whole systems of streams anastomose as do the channels of an individual braided stream within its bed; in effect, the "anabranches" of the system tributary to the Darling River in the area in question are essentially one titanic braided stream. In this horizontal country the slightest rearrangement of material, by very small tectonic† uplift or substance, or by the building of sand dunes, can produce substantial changes in the course of a stream or dam it to form a lake. Narran Lake apparently formed in some such way. The rainfall is so scanty that it never fills to overflowing, and thus has no outlet. Needless to say, its size and shape vary according to the amount of water it receives and the amount used to water livestock in this sheep-raising country. It also serves as a wildfowl refuge.

The aboriginal word *murrilla* means pebbly or stony ridges or ground. The murrilla ridges of our story are associated with "fossil" soils, developed at a time when the climate was less arid, with alternating wet and dry seasons. Under those climatic conditions, a cap of silcrete (soil cemented by silica) developed in areas underlain by Cretaceous claystones, and subsequently has weathered to produce pebbly ridges.

---

* An alluvial fan is a low cone-shaped heap of material deposited by rivers issuing from mountains onto a lowland.

† Tectonic: Of, pertaining to, or designating rock structure and external forms resulting from deformation of the earth's crust. From the Greek *tekton*, a builder.

Certain natural hazards have been responsible for a number of legends, among them Scylla and Charybdis of classical mythology. These twin dangers were encountered by Odysseus on his long journey home from the Trojan War. Charybdis was a frightful gulf into which thrice a day the waters were sucked with a roar and then disgorged, creating a whirlpool which would engulf any ship approaching too near. Those who steered clear of Charybdis, however, found themselves too close to the cliff where lurked the six-headed man-eating monster, Scylla. The legendary Scylla and Charybdis were in fact the Straits of Messina (Fig. 15), the narrow passage between the toe of the Italian boot and the northeast tip

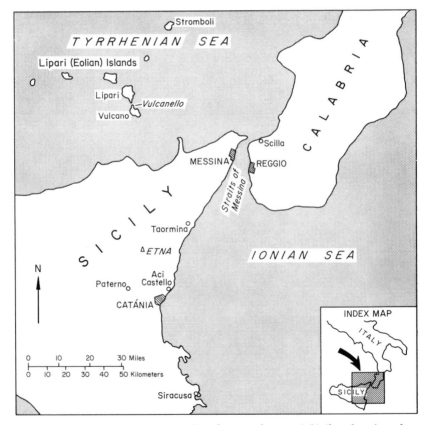

FIGURE 15. *The "toe" of the Italian boot and part of Sicily, showing the location of places associated with legends mentioned in this and other chapters.*

of Sicily. The steep rocky shore on the mainland side just north of Reggio, still called Scilla, and the dangerous currents in the straits on the Sicilian side must have posed a formidable threat to the small vessels of the ancients. The misfortunes of those lost in attempting to navigate that particular stretch of water all too naturally provided food for legend.

Another classical myth with a Sicilian setting concerns a famous spring, the Fountain of Arethusa on Ortygia, the island which constitutes the oldest part of the historic Greek city of Syracuse (Siracusa). This spring wells up to form a more or less circular pond set in a rock hollow; ducks paddle over its surface or sit preening themselves beside it, while papyrus waves its graceful fronds over its waters. It is said to be the only place outside of Egypt where papyrus grows naturally. The inhabitants of ancient Syracuse, to whom the spring was sacred, explained its presence thusly: While Arethusa, a beautiful young huntress, was bathing one day in the river Alpheus in the Peloponnesus, the god of that river became enamored of her beauty and sought to embrace her. In panic she fled, hotly pursued by the god in human shape. In desperation she called upon the goddess Artemis for help, and instantly she dissolved and turned into a spring. Alpheus thereupon resumed his fluvial form and tried to mingle his waters with hers. Arethusa sank into the ground and flowed under the Ionian Sea, emerging at last on Ortygia, but Alpheus flowed after her through the undersea tunnel and his waters after all emerged and mingled with hers. It is part of the myth that Greek flowers are said occasionally to come up from the bottom of the spring, and objects thrown into the Alpheus in Greece reappear in Arethusa's Fountain in Sicily.

Nothing in this tale, of course, has the slightest basis in fact. The source of water in springs is rain which seeps into the ground and percolates through porous strata until local geologic and hydrologic conditions permit it to flow to the surface at a particular spot. The idea of subterranean flow, however, may stem from the fact that in its upper reaches the Alpheus disappears underground for a stretch, in a limestone karst area. The Arethusa myth illustrates a very common misconception concerning ground water. When a well finds water it has not penetrated a subterranean river or pool, but a layer of rock whose pores and cracks are saturated

65

with water confined by impervious layers. The permeable layer may be a relatively shallow, unconsolidated sand or gravel deposit or it may be deeper solid rock. Only in karst areas can there be true underground streams or lakes.*

One of the landmarks of Copenhagen is the Gefion Fountain, depicting the goddess Gefion ploughing the island of Sjaelland out of Swedish soil, where it left the hole filled by Lake Vänern. There is indeed a superficial resemblance, in both shape and size, between Denmark's biggest island and Sweden's biggest lake (Fig. 16), but

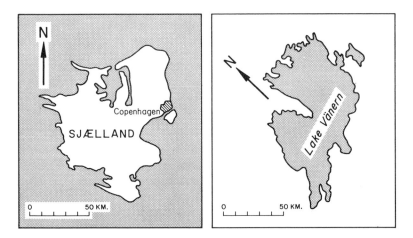

FIGURE 16. *The resemblance in size and shape between the Danish island of Sjaelland and Lake Vänern in Sweden is supposed to have inspired the myth that the goddess Gefion plowed Sjaelland out of Swedish soil, leaving the lake depression.*

that resemblance could hardly have been comprehended in the days when the myth originated, long before there were maps of Scandinavia. The myth is found in the works of Iceland's famous thirteenth-century historian Snorri Sturlusson, who in turn quotes

* The same misconception applies even more to petroleum, where the use of the terms *oil pool* and *reservoir* by the industry does nothing to dispel the notion. Actually, when an oil geologist speaks of an oil pool he is thinking of a layer of rock whose pores are saturated with oil; when that layer is penetrated by the drill, the oil seeps into the well and either flows unaided to the surface or is pumped up, depending on the pressure, the pressure in turn depending on the geologic and hydrologic setting. An *oil sand* is usually a sandstone, not a loose sand. A porous standstone makes an excellent reservoir rock for either oil or ground water.

a verse in which it is mentioned by the poet Bragi Boddason, who lived in the first half of the ninth century [241]. In the early versions the lake in question is Lake Mälaren. Snorri's version runs something like this: It is told that in return for her entertaining him, King Gylfi of Sweden offered a woman traveler as much land as could be ploughed in a day and a night by four oxen. Unbeknownst to him, the woman was Gefion in disguise. From the abode of the giants she brought in four oxen, her own sons by a giant. She yoked them to the plough and ploughed so deep and wide a furrow that a portion of Gylfi's land was torn away. This she had the oxen haul to the sea, where she placed it and named it "Sea Land" (Sjaelland, in Danish). Where she had taken the land a lake was left, called "Lögr" (Mälaren), whose inlets correspond to the headlands of Sjaelland. The myth may have been transferred to Lake Vänern after maps were drawn, revealing the closer resemblance of Sjaelland to the latter.

Another kind of continuing natural phenomenon has given rise to geomyths. Natural earth fires burning near the old Lycian port of Phaselis, where natural gases seeped out of the ground, may have been the origin of the fire-breathing chimera slain by Bellerophon [107]. In fact, it has been suggested [107] that Prometheus, who gave fire to mankind, and Hephaistos, who in earlier myths is a kindly and peace-loving deity who used fire for the benefit of gods and men, may both have been personifications of the power of fire offered by Nature herself. They may have originated in Asia Minor or the Caucasus and been transplanted to Greece later. On Lemnos, a favorite abode of Hephaistos and sacred to him, there was a hill called Mosychlos from which natural fires once issued, but the hydrocarbon gases which fed the flames have long since been exhausted [107].

If large-scale landforms are to be considered the work of giants in or on the earth, then it would be logical to attribute certain small-scale natural features to the work of little creatures. A rather interesting suggestion has been made regarding the possible origin of the notion of the dwarfs in Scandinavian mythology [91:242]. The Phoenicians are known to have mined iron, copper, gold, and tin as far away as England, Norway, and Sweden. They also were very secretive about the location of their mines. Might they have played on the credulity of some of the early inhabitants of those areas by deliberately fostering the belief that their miners were a supernatural race who dwelt underground? Although the idea of the little creatures is implicit in the dwarfs of Scandinavian mythology, at least in the Wagnerian concept of them as the Nibelungen busily laying up golden treasure deep in the earth, further specific examples are not easy to find.

However, folklore in many other forms, from myths to current superstitions, is firmly associated with various individual minerals or with concentrations of minerals into useful deposits of ore. From earliest times gems and precious stones have been esteemed for their beauty and rarity; they also have been valued as charms to protect the wearer against various misfortunes or to bring misfortune to his enemies. It has been suggested [133:1] that the use of precious stones as jewelry stems from their original use as amulets and talismans. A trace of this notion survives in the custom of wearing one's birthstone. All gems and precious stones, and many semiprecious ones as well, are surrounded by a wealth of folklore and superstition. In some cases there is a fairly obvious relationship between the supposedly supernatural properties of the stone and some individual physical property, like the hardness of the diamond for instance. In others the reason for the superstition attached to a particular stone is obscure, there being no relation between the power attributed to the stone and any intrinsic property of the mineral. The folklore of gems and minerals is so extensive that we will limit ourselves here exclusively to cases where the superstition or legend associated with the stone is occasioned by its inherent nature as a mineral.

In our time the opal is almost as firmly associated with bad luck as the number **13** or black cats, but it was not always so. Before the nineteenth century many virtues were ascribed to the opal, including the capacity to protect its wearer against disease. One reason for the present superstition may be its fragility. Opal is hydrous silica, in a form which behaves to x-rays as though it were amorphous—that is, as if its molecules are not arranged in any ordered crystal structure. The play of color which characterizes the precious varieties of opal is caused by the breaking up of light, but the exact cause of that breakup was not definitely known until very recently. Now the electron microscope has revealed that there is some order to the structure of opal after all. Spherical particles less than a micron in size (a micron is one-thousandth of a millimeter) form what crystallographers call face-centered cubic arrays (Fig. 17), and these diffract the light [87].

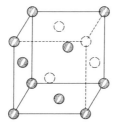

FIGURE 17. *Face-centered cubic crystal structure. A rudimentary submicroscopic structure of this type is responsible for the play of color of opal.*

Previously it was thought that the microscopic and submicroscopic cracks in opal, caused by strains set up during the drying out of the original silica gel, refracted the light to produce the interference colors. Those internal strains might be a source of bad luck —but not to the wearer. Sometimes opals shatter upon being brought up out of the ground, and that is certainly bad luck to those who hope to make a profit of mining them. They shatter very easily while being cut and polished—hard luck for the lapidary entrusted with the job. Once fashioned into wearable gems, opals must be handled with care; they shatter easily if roughly treated (but so do many other stones), and can absorb moisture or oils along flaws. The finest opals show some loss of life and color after a century or so, and poor ones within a few years. But with gentle treatment a fine opal will give its wearer pleasure for a lifetime, and after all, isn't that good luck of a sort?

"Fairy stones" are twinned crystals of staurolite, natural stones in the shape of a cross. Staurolite, a silicate mineral, develops in certain shaly rocks when they are metamorphosed into schists under the action of heat and pressure. Staurolite has a pleasant reddish-brown color and is somewhat glassy in luster; it often develops rather perfect crystal form and is easily removed from its matrix. Occasionally a transparent crystal of staurolite is cut for a gem, but most staurolite is opaque and would not be attractive enough to use as jewelry were it not for its distinctive twinning. Twin crystals are the intergrowth of two or more individual crystals in such a way that certain of their crystal planes (which reflect the layering of the atoms) are parallel while others are reversed with respect to each other. In staurolite (the name comes from the Greek *stauros*, meaning cross) the twins are more or less perfectly cruciform, the two individuals interpenetrating either at right angles or at a sixty degree angle, depending on which crystal plane is the twinning plane (Plate 15). The natural occurrence of perfect crosses in a

PLATE 15. *Staurolite crystals. Left: An untwinned crystal. Center and right: "Fairy Stones"—cruciform twinned crystals (90° and 120°, respectively). (Photo by courtesy of L. G. Berry.)*

rock inevitably has been associated with the crucifixion. In Brittany staurolite twins were thought to have been dropped from heaven, and were prized as charms. Fine staurolite twin crystals are found in Patrick County, Virginia. The legend [133:271] there has it that long ago a band of fairies were dancing around a local spring when a messenger arrived bearing news of the crucifixion far away. The fairies wept to hear of Jesus' suffering, and their tears crystallized in the shape of a cross. (With its mixture of fairies and Christian elements, this legend smacks of commercial "fakelore," I fear.)

Along the coasts of Hawaii one frequently finds boulders of

lava rock which contain small pebbles or grains of sand embedded in rounded cavities; these are often so firmly fixed that it is difficult to imagine that they could have been introduced after the lava hardened. The Hawaiians call them *hanau* or "birth stones," and claim that they are babies about to be born to the larger stone (Plate 16). The real explanation seems fairly simple. The upper

PLATE 16. Hanau *or birth stones, Ninole Cove, Hawaii. The Hawaiians believed that the small pebbles lodged in the vesicles of basalt boulders were babies about to be born to the larger stones. (Photo by the author, August 1970.)*

part of lava flows commonly contain cavities (vesicles) representing bubbles of gas that did not have time to escape before the rock congealed. As chunks of such rock are rolled and pounded and smoothed by the waves, the vesicles are hollowed out and enlarged, and generally smoothed too. Storm waves pile some of the boulders up on the beach, where they are subsequently pelted by sand grains, pebbles, and larger fragments carried by the incessant surf. Sooner or later, some sand grain or pebble of just the right size strikes a vesicle of just the right shape, at just the right angle, with such force that it becomes firmly wedged in, and voilà—a *hanau*. Further erosion by the sandblasting action of waves may loosen the *hanau* in their setting, whereupon, if they are not washed out again, they may churn around and enlarge their hollows to form miniature potholes; I have seen some over an inch deep and containing several pebbles. *Hanau* can also be found in blocks of lava used to build sea walls.

Over a vast area in the southern part of Australia, in the soil or weathered out on the surface, there have been found thousands of small, black, glassy bodies of remarkable shape, unlike any "normal" geologic object. These stones, which have been given the name *australites*, are seldom more than one or two inches across, and are frequently much smaller. Their shapes vary considerably, the commonest being round in plane and lens-shaped in side view, sometimes distinctly flanged; some are oval, while others are shaped like dumbbells, teardrops, or boats or canoes (Plate 17). The sharp-eyed aborigines recognized these stones as unusual and regarded them with superstitious awe as objects of magic and mystery [6]. Not unnaturally, they speculated as to the origin of australites. Some considered them to be emu eyes lost while the birds were

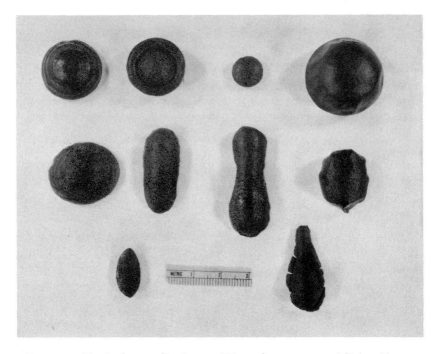

PLATE 17. *Typical australite forms. (Photo by courtesy of Brian Mason, Smithsonian Institution.)*

foraging for food, or "staring-eyes" belonging to ancestral beings. Others thought they were produced by lightning, since they sometimes were found in the soil at the roots of trees struck by lightning. Still others thought they fell from the sky. The staring-eye or emu-eye explanations, based purely on the physical similarity of the round australites to such eyes, have no more resemblance to the facts than the whimsical name of "blackfellows buttons," which the early settlers gave to the australites in joking reference to the fact that the customary attire of the aborigines consisted of the absolute minimum. The lightning explanation is not too bad a guess, for lightning actually can melt rock where it strikes—but the resulting "fulgurites" (named from the Latin for "thunderbolt") are irregular tube-shaped bodies, not at all like australites. The last explanation, that they fell from the sky, happens to be correct, but just by chance, inasmuch as the fall of the australites, though geologically recent, predates the coming of man to Australia.

The australites are a kind of *tektite* (from the Greek *tektos*, melted). Tektites are found not only in Australia but in several widely separated areas of the earth's surface, and are named according to the site of the occurrence. The indochinites, from Indochina and Thailand, are also geologically recent and may be part of the same fall; the Ivory Coast tektites are about a million or so years old; the moldavites from Czechoslovakia (named for the Moldau River) are somewhere between 13.5 and 20 million years old; and the bediasites from Texas (named for the Bedias sandstone with which they are associated) and a few tektites found in Georgia, presumably belonging to the same fall, are about 34 million years old.

Some of the names given to tektites in parts of the world other than Australia reflect local ideas concerning their origin [7]. On the island of Hainan in the South China Sea they were known as "star excrement," "devil's droppings," or "moon stones." Other Indo-Malaysian tektites have been called "thunder dung," "sunstones," "moonballs," and "devil balls." Everywhere, they have commonly been credited with magic powers, if only as good luck charms. Even contemporary Australian gold miners, like the natives of the Ivory Coast, have a superstition to the effect that their presence in stream gravels is a sign of a rich placer gold deposit.

The origin of tektites is currently the subject of much debate.

They definitely are not a kind of meteorite, as first thought, but they do appear to have been detached from some body by the impact of a meteorite or comet. The question has been, was that body the earth, or the moon? The absence of the isotope aluminum-26 shows that they have not been in space long enough to have come from farther away. Evidence for a terrestrial origin rests mainly in the chemical and isotopic compositions of the tektites (which closely resemble those of typical rocks or soils of the earth's crust), in the grouping of tektite localities, and in the fact that the molda-vites seem to be related to the Ries impact crater in Germany and the Ivory Coast tektites to the Bosumtwi meteor crater in Ghana. However, aerodynamic analysis of the shape of the tektites and physical and chemical differences between tektites and terrestrial glasses definitely known to have been formed by meteor impact favor an extraterrestrial origin. The moon's gravity is low enough that material thrown up upon impact of a large meteorite could easily escape into space, find its way into an earth orbit, and become melted upon entry into the earth's atmosphere; but the chemical composition of the moon samples returned by the Apollo missions has not given comfort to those who support a lunar origin of tektites.

Unlike tektites, which, however interesting from a scientific point of view, have no immediate economic value other than as museum specimens or curios, mineral deposits are of great practical interest. Their origin has been the subject of speculation from earliest days. The Ngadjuri tribe in South Australia, for instance, has a tale [195:15] which explains how two deposits of mineral pigments, with which they decorate themselves for rituals, came to be. From somewhere to the northwest there came an old woman and her two savage dogs, one red and one black. The old woman was a cannibal, and the dogs would kill people for her and share in the feast. People often had to abandon their camping grounds to keep out of her way. When word came that she was approaching one of the main camping grounds, however, the people decided to try to kill the savage trio rather than flee. Two brothers were chosen to do the job, and taking their boomerangs they set out to meet them. One of the brothers, hidden in a tree, called to attract attention. The red dog spied him and made for the tree, whereupon the other brother stepped out of his hiding place behind a bush and threw his boomerang with such skill that it cut the dog in two. Again the tree-climber called out, and the black dog was similarly served. Then the brothers killed the cannibal woman, and the threat was ended. Where the red dog's blood had been spilled, a deposit of red ochre was created, and where the black dog's blood flowed, a deposit of black ochre. (Red ochre is an impure powdery form of the iron oxide mineral, hematite; black ochre, or wad, is an impure mixture of manganese oxide and other oxides.)

The discovery of gold or silver is in itself a glamorous event, and even within our memory, tales which are often more folklore than fact have sprung up around the finding of particular deposits. The discovery of silver at Tonopah, Nevada, in 1902 is credited to a donkey, which is said to have kicked up a piece of ore and thus drawn it to the attention of his master, prospector Jim Butler.

When the beginnings of a mining enterprise are lost in the mists of antiquity, the circumstances surrounding the initial discovery acquire the patina of true legend. Such is the story of Banská Štiavnica, an old Slovakian mining town, which, according to Tacitus, was already producing gold and silver in the first century A.D. (The gold and silver have long since been exhausted, but the Banská Štiavnica deposit is still being mined for lead, zinc, and copper.) A certain man, so runs the legend, had two salamanders, one endowed with the ability to smell out gold and the other silver. All he had to do was to turn them loose, follow where they led, and if there was gold or silver in the ground they would indicate where to dig. These salamanders, accordingly, showed him where the Banská Štiavnica ore was to be found. So firmly is the legend attached to this ore deposit that the insignia of Banská Štiavnica shows two salamanders, and holiday processions there are led by a man bearing the figure of an exaggeratedly large salamander. There is no way of knowing whether a salamander really had anything to do with the discovery. One might speculate that perhaps somebody saw a salamander disappear down a hole or under a rock—a perfectly natural thing for a startled creature to do—and, inasmuch as salamanders were long credited with magical powers (including the ability to live in fire), he might have been tempted to investigate and—Eureka!

75

Going back still further into the past we come to Jason of classical mythology, whose main adventure, the capture of the Golden Fleece, has geologic overtones. Jason, the son of a king of Thessaly, set out with his band of Argonauts to capture the Golden Fleece hanging in a sacred grove in the kingdom of Colchis on the Euxine (Black Sea), where it was guarded by a sleepless dragon. Aided by the sorceress Medea, princess of Colchis, Jason performed prodigious feats and finally succeeded in seizing the Golden Fleece and escaping with it, taking Medea back to Thessaly with him. An early interpretation of this myth is that the Argonauts' expedition was semi-piratical, the Golden Fleece representing the booty they brought back. A more interesting possibility from the geologic point of view has been proposed by T. A. Rickard [203:1047]. In his opinion, Jason was the spiritual ancestor of the Forty-Niners of California, the Sourdoughs of the Klondike, and all gold seekers through the ages. It seems there was a tribe called the Tibareni in ancient Colchis who practiced a form of placer mining by sluicing gold-bearing stream gravel over unscoured sheepskins, which caught and held the particles of gold. After shaking out the coarser flakes and nuggets, they hung the fleeces on trees to dry, and then beat out the fine gold dust. It was the rumor of these "golden fleeces" that lured Jason's expedition to Colchis. Also implicit in this explanation, claims Rickard, is the modern metallurgical method of flotation*—the natural oil in the fleeces caught and held the metallic particles.

Pactolus is the name of a tiny defunct mining camp in the Paradise Range of Nevada, where a hoped-for gold strike never materialized. It was obviously named by someone with a classical education. The Pactolus River in Lydia, Asia Minor, was a source of placer gold in ancient times. The geologic explanation of how the gold came to be in the river is prosaic compared to the myth which inspired the name of the Nevada mining camp. Placer gold deposits are formed when rocks containing gold veins are eroded, and the heavy gold particles become concentrated in the stream bed

---

* Flotation: A method of separating the different minerals in pulverized ore according to their ability to float in a frothy liquid; the finely ground ore is first treated with an oily substance, which enhances the difference in wettability between the metallic and nonmetallic particles.

alluvium by the action of running water. Sometimes such deposits can be traced upstream to the source rocks, but sometimes the lode has been completely eroded away. This may have been the case with the Pactolus River, for one of the most picturesque of myths sought to account for the presence of gold in the river, the myth of King Midas, who was granted the wish that everything he touched might turn to gold. Midas quickly discovered that the "golden touch" was a mixed blessing, to say the least; all his food and drink turned to gold the instant it touched his lips. He prayed to be cleansed of the power and was directed to bathe in the source of the Pactolus. The power was thereby transferred to the stream, whose sands turned to gold.

Up to this point we have been considering folklore which has been or could have been inspired by some geologic fact. In closing this chapter, let us consider an example of "factlore" with a mineralogical basis. Long before the magnetic compass was known to the Vikings—it appeared in Scandinavia after about A.D. 1200—they were navigating the open waters of the North Atlantic with great precision, ranging as far as the continent of North America. How were they able to keep on course on cloudy days, which must have been frequent at certain times of the year? According to the sagas, they used an extraordinary stone called the *solarsteinn,* or sunstone, with which they could ascertain the sun's direction even when the sky was completely overcast. Although it was obvious that they must have had some sort of navigation aid, the idea of a direction-finding stone was long dismissed as nothing but folklore.

Recently it has been suggested by Dr. Thorkild Ramskou [193] not only that the sunstone was a reality, but that its basic principle was the same as that used in the modern twilight compass. That instrument, invented in 1948 for the United States Navy, is used by pilots who fly the North Pole route. (The magnetic compass, of course, becomes erratic and unreliable in the vicinity of the earth's magnetic pole.) At dawn or sunset, when the sun is low in the sky, its rays impinge horizontally upon the upper atmosphere and are reflected toward the ground. Reflected light is always polarized—that is, the light rays vibrate in a single plane rather than in all directions around the path of propagation. As anyone knows who has ever played with two pieces of Polaroid, when a polarized beam of light is viewed through another polarizer the amount of light that passes through the second one varies according to its orientation with respect to the first: when both planes of light vibration coincide, light passing the first polarizer also passes unhindered through the second; but when the planes are at right angles to each other, no light can get through the second (Fig. 18).

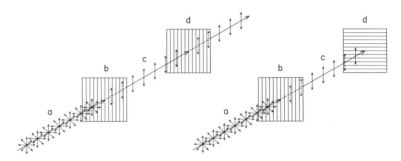

FIGURE 18. *Polarization of light. Left: unpolarized light (a), vibrating in all directions around the direction of propagation, passes through a polarizing substance (b); only rays (c) vibrating in the same plane as the polarizer are transmitted, and they are not obstructed by a second polarizer (d) whose plane of polarization is parallel to that of (b). Right: the polarized beam (c) emerging from (b) is completely stopped by a second polarizer (d) at right angles to the first. For intermediate positions of (d), more or less light will be transmitted, depending on its position with respect to (b). Light also becomes polarized when it is reflected.*

Thus, if there had been just enough break in the cloud cover overhead to show a patch of blue, the Vikings could have determined which way was east or west if they had some sort of stone which was sensitive to the direction of polarization of light reflected from the sky (Fig. 19). Of course the sagas must exaggerate when they say the sunstone worked even when the sky was completely overcast, if it worked on the principle described.

What mineral could the Norsemen of a thousand years ago have used as their sunstone? Iceland spar comes first to mind, because it is the mineral that polarizes light most efficiently due to its extremely high double refraction (birefringence). It is a very clear variety of calcite, used in the petrographic (polarizing) microscope and other optical instruments. But the crystals must be cut at a mathematically precise angle and prepared in a very special way in

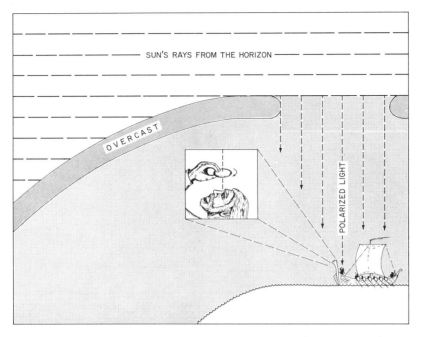

SUN'S RAYS FROM THE HORIZON

OVERCAST

POLARIZED LIGHT

FIGURE 19. *The principle of the Vikings' sunstone. (After Ramskou, 1967* [*193*].)

order to reflect out one of the two refracted rays and transmit only the other, polarized in a single plane; the Vikings certainly were not technologically sophisticated enough to have discovered how to make that kind of polarizer out of Iceland spar.

Ramskou therefore has suggested that they used crystals of cordierite (dichroite) or andalusite, both of which can be found in beach gravels along the Norwegian coast where they have weathered out of the metamorphic rocks in which they occur. Neither of these minerals is as strongly birefringent as calcite, but they exhibit to a very marked degree another effect known as *pleochroism*. Owing to their crystal structure, they absorb light more strongly in certain directions than in others. When viewed against polarized light and rotated, they show a definite change in color. The thicker the crystal, the more noticeable the change. (Colorless minerals, obviously, cannot be pleochroic.) The Vikings, thinks Ramskou, could have observed and utilized the pleochroism of cordierite or andalusite to determine the direction of the source of polarized light.

While crystals of cordierite or andalusite of high quality (transparency) might have been found by the Vikings, mineralogists object to this explanation because crystals *large* enough to be effective as sunstones are unknown from Scandinavia. And while it is true that a perfectly clear cleavage piece of colorless calcite will show no change in the amount of polarized light transmitted in different directions (because the two rays refracted through it are polarized at right angles to each other, and the more one is cut out the more the other comes through, allowing the same total amount of light to pass through the crystal in any direction), there is a way in which calcite could have served as the sunstone. It has been demonstrated at the Mineralogical Museum in Copenhagen [181] that if polarized light enters a piece of clear calcite through a fine grating, there is a noticeable change in the amount of light transmitted in different directions. The Museum demonstration uses as the grating a piece of paper with two slits forming a ninety degree cross, but the same effect can be achieved with a piece of calcite, one surface of which is a naturally etched crystal face rather than a shiny cleavage face. It is entirely possible that such a piece once came into the hands of some Viking who accidentally discovered its unique properties. Suitable pieces of calcite would not be common—the sagas mention that they were very costly—but they would be very much easier to find than large gem-quality crystals of cordierite or andalusite. The sagas also tell how a thief once threw away a sunstone thinking it was a piece of quartz. If the sunstone was a colored mineral, it seems very unlikely that any thief would have thrown it away even if he took it for quartz, for the clear colored varieties of quartz are semiprecious stones in their own right. To an untrained eye, however, colorless calcite could quite easily be taken for common colorless quartz, a fact which lends additional credence to the idea that the sunstone was Iceland spar.

5 /

Earthquake Lore

> ... some say the earth
> Was feverous and did shake.
>
> MACBETH

EARTHQUAKE LORE SEEMS TO BE ALMOST ENTIRELY etiological. I have been able to locate only two examples of euhemeristic legends which can be attributed to specific earthquakes: one, concerning the New Madrid earthquake of 1811, is related in this chapter, while the other, being part of the Araucanian Indian flood tradition, will be found in the chapter on the Deluge. Even in those parts of the world which are not particularly seismic, attempts have been made by the inhabitants to explain why the ground occasionally trembles. In his book *Causes of Catastrophe* [139:11ff.], L. Don Leet rounds up more than twenty such explanations from all over the world, and most of those quoted below which are not otherwise credited are from that source. The most prevalent concept involves some creature or divinity residing in or under the earth, and were it not for the fact that geologic misconceptions are unusually well represented in connection with earthquake phenomena, this chapter would consist largely of variations on a single theme.

Very commonly the creature responsible for earthquakes has been thought of as supporting the earth. The Algonquin Indians pictured the earth as being carried on the back of a giant tortoise. In the Celebes they believed it was a huge hog, which caused tremors when it scratched itself against a palm tree. In the Moluccas and Sumatra the supporting animal was a serpent, and in Persia a crab. In places as far apart as Bali, Borneo, Bulgaria, Malaya, and Constantinople it was thought to be that beast of all work, the buffalo. Lamas of Mongolia thought the earth rested on the back of a gigantic frog, and that earthquakes occurred directly above whatever part of its body it twitched. In the Brahman mythology of India the seven serpents who guard the seven segments of the lowest heaven take turns supporting the earth; quakes occur when they change shifts. In other Indian mythology, eight elephants hold up the earth, which shudders when one of them tires and shakes his head. According to the Indians of southern California, the earth was held up by seven giants.

The Tlascaltans of Mexico believed that the earth rested on the shoulders of certain divine beings, who caused it to quake as they shifted it from one shoulder to another. In Latvia long ago it was believed that a god named Drebkhuls carried the earth around with him and jarred it as he walked. The inhabitants of Nias, one of the islands of Indonesia, thought the earth was upheld by the demon Ba Ouvando, who shook it in rage if they failed to make the proper sacrifices to him. In the Manichean religion the earth was said to be supported by the giant Homophore; earthquakes occurred when he fought with another giant. In Colombia the god Chibchacum was condemned to carry the earth on his shoulders as punishment for having flooded the plain of Bogota as a prank; before that it had rested on a firmer foundation in the form of three beams.

In the part of West Africa formerly known as Senegambia, the aborigines noted that earthquakes always seem to come from the west, and explained them thus: After its creation, the earth was placed on the head of a giant. All the things that grow in the earth are his hair, and all the creatures that move on the earth are the parasites crawling in that hair. This giant sits facing the east, but occasionally he turns quietly to face the west; then when he turns back toward the east he does so clumsily, and jars things loose on top of his head. (The African continent as a whole is seismically stable; even in the rift valleys of East Africa and the Red Sea, earthquake activity is moderate compared to other active belts. But like most other stable masses, Africa is rimmed by marginal fractures which are seismically active. Thus in West Africa, shocks would come from the west.)

Another imaginative African explanation of earthquakes is the belief of the Wanyamwasi tribe that one side of the earth's disk rests on a mountain, while the other is held up by a giant, whose wife in turn holds up the sky. Whenever the giant embraces his wife, the earth trembles.*

An inanimate but nevertheless precarious support for the earth, envisioned in a Christian legend from Rumania, is that it rests on the three pillars Faith, Hope, and Charity; if one of these is deficient on earth, the corresponding pillar contracts and the earth teeters out of balance until God restores its equilibrium. (One defect of this legend is that the earth could be very gravely deficient in all three virtues equally, and yet be in no danger of earthquake!)

Most such legends neglect to mention what supports whoever or whatever supports the earth, but there are a number which picture an aquatic creature swimming in some sort of universal ocean. The Masawahili tribe of East Africa dreamed up a complicated sort of circus act in which a giant fish carries on its back a stone, on which stands a cow balancing the earth on one horn; an earthquake means that the cow has shifted the burden to the other horn to ease its aching neck.

The most elaborate earthquake folklore comes, not surprisingly, from highly seismic Japan. The current superstition attributes earthquakes to the wriggling of a giant catfish (*namazu*) under the earth, whose head is under the province of Hitachi. His movements are kept under control as much as possible by the Kashima deity, by means of the *kaname-ishi,* or "pivot-stone" (Plate 18). Exactly when the *namazu* legend originated is not certain. According to C. Ouwehand [173:52-53], it probably developed from the idea of a serpent-dragon surrounding Japan, representing (as it does in some

---

* When I quoted this legend in answer to a facetious query from a colleague as to whether there was any sex in my book, it provided him with an opportunity to get off a little "in" joke: "That must be the origin of Love waves." Love waves, a form of surface wave produced in earthquakes, are named after the British mathematician A. E. H. Love, who did some of the fundamental theoretical research on seismic wave propagation.

PLATE 18. *To prevent earthquakes, the Kashima deity pins down the Namazu (catfish) by means of the "pivot stone," while the people pray. The characters read "Kaname-ishi," meaning pivot stone. (Drawn by Kenzo Yagi after one of the old Namazu-e prints.)*

other cultures) the primeval sea surrounding and supporting the earth. Mutually interchangeable concepts of this serpent-like creature (almost insect- or spider-like in some representations), of a giant fish or whale, and finally of the catfish retained the basic underlying concept. Then in the last decades of the seventeenth century the association of the Kashima deity and pivot stone with the earthquake *namazu* appeared in pictorial representations, later to become crystallized in the current form as a result of a series of prints issued soon after the Edo* earthquake of 1855, the "Ansei era earthquake disaster." It was not the severest earthquake of that

* The old name for Tokyo.

era, but as Ouwehand puts it, "it seems, however, after the opening of the Ansei era—so turbulent, politically as well—with natural disasters and famine in several parts of Japan, to have unquestionably been the event that aroused the most intense appeal to the imagination" [173:3]. The appearance of the *namazu* prints was a reaction to that catastrophe. In addition to representing the event and several versions of the *namazu* myth, the pictures and texts of these typically folk prints satirized certain social conditions. The *how* of the earthquake is explained by the Kashima deity's absence, exploited by the *namazu;* the *why* is punishment for social abuses— and also a result of the visits of Commodore Perry's squadron in 1853 and 1854! The satire expressed in the prints was intended "to make life more livable for the masses whose existence, even before the earthquake, was already far from enviable" [173:22]. The prints were also bought as charms against subsequent earthquakes.

In some earthquake legends the shocks are attributed to creatures within, rather than supporting, the earth. In one part of India there was thought to be a giant mole whose burrowing shook the earth above. The Kukis of Assam may still believe in a race living inside the earth, who shake the ground to find out if anyone still lives topside. Whenever the Kukis feel a tremor they call out "Alive, alive!" to assure them that someone does. The Karens of Burma blamed earthquakes on the activities of a god Shie-Ou, imprisoned in the earth by the sun god Ta-Ywa [139:15]. The Tongans attributed earthquakes to a god, reposing in the volcano which forms the island of Tofua, who stirred or turned over in his sleep. The old Norse mythology also attributed earthquakes to a demon in the earth: for his misdeeds Loki, the demon or demigod of malice and mischief, has been imprisoned by the gods in a cave, chained in such a way that he lies on his back on three sharp stones; above his head hangs a poisonous snake, its venom constantly dripping on his face. Loki's faithful wife Sigyn stands beside him catching the poison in a bowl, but occasionally she must leave his side to empty the bowl and the poison falls on his face, whereupon he gives a violent start which shakes the whole earth [91:227].

Another type of earthquake legend attributes shocks to the heavy tread or stamping of some giant or deity. For instance, the Basoga tribe, who live on the north shore of Lake Victoria Nyanza in Africa, believe in an earthquake god named Kitaba who shakes the earth if he walks too fast [139:15].

Still other earthquake lore personifies the earth. The Kaffirs in Mozambique think of the earth as shaking in the grip of an ague [139:16], an analogy which Aristotle used figuratively, as did Shakespeare in the quotation at the head of this chapter. From Peru comes the more cheerful thought that the earth occasionally kicks up its heels and dances [139:17].

In Greek mythology, Poseidon was the Earth Shaker. Apparently the ancient Greeks recognized the difference between the very local volcanic earthquakes (which they attributed to the struggles of imprisoned giants) and tectonic earthquakes. Although it might appear strange at first glance that they did not blame the latter on Atlas, who supported the earth, their choice of the sea god is not really surprising if we consider how many of the shocks experienced in Greece and its islands originate under the sea and how often they are accompanied by tsunamis,* large or small. Later, the Greek philosophers speculated seriously about the cause of earthquakes, but their explanations were almost as wide of the mark as the primitive legends. According to Aristotle earthquakes were caused by winds struggling to escape from imprisonment in subterranean caverns. This idea persisted at least until Shakespeare's time, for in *Henry IV*, Part I, Hotspur explains to Glendower how

> Diseased nature often times breaks forth
> In strange eruptions; oft the teeming earth
> Is with a kind of colic pinched and vex'd
> By the imprisoning of unruly winds
> Within her womb; which for enlargement striving
> Shakes the old beldam earth, and topples down
> Steeples and moss-grown towers.

In the Old Testament earthquakes were regarded as signs of divine wrath, and the Church of the Middle Ages continued to emphasize the idea of punishment for failure to keep in line. In this view, naturally, no scientific mechanical explanation of earthquakes

---

* *Tsunami* is a Japanese word used internationally as the scientific term for what otherwise are known as seismic sea waves, which are popularly but erroneously called "tidal waves." Much more will be said about tsunamis in later chapters.

was necessary, but a pseudo-scientific one was eventually offered nonetheless in 1682 by J. B. Van Helmont. Van Helmont was a Belgian chemist and physician who was a curious mixture of mystic and alchemist and sound scientist; he was the first to understand the nature of gases as distinct from air, and made contributions to our understanding of nutrition and digestion. His explanation of earthquakes was that an avenging angel struck the air so as to give rise to a musical tone, whose vibrations were communicated to the earth in a series of shocks [139:20].

In this country, earthquakes undoubtedly have occasioned more than a few hellfire-and-damnation sermons. It is said that after the New Madrid earthquake of December 16, 1811, whose aftershocks continued for years, many a repentant sinner returned to the fold. The New Madrid earthquake may well have been the strongest earthquake ever experienced in North America, but the area was rather sparsely settled at the time and of course there were not yet any seismographs to obtain pertinent scientific information at a distance [47]. One of the few euhemeristic earthquake legends concerns the reason for the New Madrid earthquake and some of its topographic results.

According to this legend, there was a handsome Chickasaw chief named Reelfoot, who unfortunately was born with a club foot. Because of Reelfoot's deformity, the father of the beautiful Choctaw princess whom he loved refused to entertain his suit for her hand. Not easily discouraged, Reelfoot and his friends carried off the princess and proceeded with the wedding, thereby incurring the anger of the Great Spirit. In the midst of the festivities the Great Spirit stamped his foot and caused the earth to tremble. The Father of Waters (the Mississippi River) reversed his course and flowed overland and submerged Reelfoot, his bride, and the whole wedding party under the waters of a new lake [35].

Reelfoot Lake, on the Tennessee side of the Mississippi, was in fact created as a consequence of the New Madrid earthquake (Fig. 20) [65; 110:10ff.]. A swampy tract sank several feet and subsequently filled with water. The Mississippi did appear to flow backwards for a short while after one of the first and severest of the shocks; this could have been due to temporary blocking of the channel by landslides or to disturbances caused by uplift of bedrock areas in the vicinity [111:9]. The depression occupied by

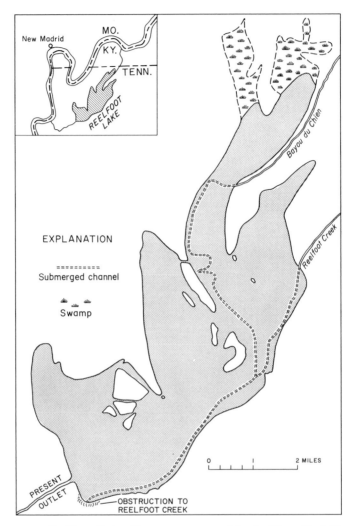

FIGURE 20. *Reelfoot Lake, Tennessee, created in the New Madrid earth-quake of 1811. (After Fuller, 1912 [65].)*

Reelfoot Lake might have been filled suddenly by waves overflowing the banks of the river to some distance, or slowly by the creeks draining into the original swamp.

I have been unable to ascertain whether this is an authentic Indian legend or just another example of fakelore. However, there is another tradition of disaster which possibly is based in part on an earthquake and which is indubitably authentic—the destruction of Sodom and Gomorrah. Genesis tells us that Sodom and Gomorrah and two others of the five "Cities of the Plain" (Admah and Zeboim) were destroyed in punishment for their wickedness by a rain of fire and brimstone from heaven. The ruins of these cities have never been found. According to J. P. Harland [98], the evidence from the Bible and from later Greek and Latin writers indicates that they must have been located in a fertile area around the southern end of the Dead Sea. Since the level of the Dead Sea has risen in past centuries, the area, the biblical Vale of Siddim, is now submerged (Fig. 21); islands which were described in the nineteenth century are no longer visible, and the shoreline has shifted progressively southward. An early suggestion as to a possible nat-

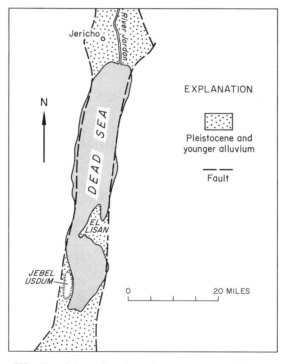

FIGURE 21. *The Dead Sea, showing places mentioned in connection with the destruction of Sodom and Gomorrah. The "Cities of the Plain" were probably located in the area now covered by the waters of the southern embayment. Jebel Usdum is a salt dome, from which a number of pillars of salt have been carved out by erosion in the course of the centuries. (After Clapp, 1936 [34].)*

ural cause of the disaster that overtook Sodom and Gomorrah was that of a volcanic eruption, but the geologic evidence rules that out. The destruction of the Cities of the Plain is thought to have occurred about 2000 B.C., and there are no volcanic rocks as young as that in the area.

A geologically more plausible explanation has been suggested by F. G. Clapp [34]: The Dead Sea lies in a graben, or rift valley, and the region is highly seismic because movement along the boundary faults is still going on. Once the Dead Sea was known as "Lake Asphaltites," because of the masses of bitumen that rise to the surface from time to time. These masses apparently are squeezed out of seepages under the water and are particularly noticeable after earthquakes. Masses as big as houses have been reported. This floating bitumen is collected and sold by the local inhabitants, and judging from the widespread trade in Dead Sea bitumen among the ancients, the deposits may have been more extensive in the past. The numerous "slime pits" of the Vale of Siddim were such seepages, probably dug out to some extent to obtain the bitumen. In addition to this natural asphalt, the region abounds in bituminous rocks, some containing such a high percentage of bitumen that they will burn. There are also a few oil seepages. Ancient writers also mentioned foul odors emanating from the Dead Sea waters, and an "invisible soot" that would tarnish metals—presumably sulfur gases; these are no longer observed, but natural gases are normal associates of petroleum manifestations and would be the first to escape completely from the ground. All these phenomena—natural gas, petroleum, and bitumens—are in this case related to the intrusion of a large mass of rock salt, or salt dome, the hill called Jebel Usdum (see Fig. 21).

In such a setting all that is required to produce a holocaust is some agency, natural or human, to set the combustible material alight. In view of the biblical statement that it was fire from heaven, lightning immediately suggests itself. Brimstone, of course, is just another name for sulfur, and the penetrating odor of burning sulfur could easily have been noticeable among the other smells of burning. But it does not seem likely that a bolt of lightning alone could have produced a fire so uncontrollable that it devoured four separate cities. Putting together ideas advanced by Clapp [34] and Harland [98] and others quoted by them, this is the sort of picture that emerges: A disastrous earthquake shook the Vale of Siddim in about 2000 B.C., releasing large quantities of natural gases and bitumens which were ignited by scattered hearth fires. The resulting conflagration wiped out Sodom, Gomorrah, Admah, and Zeboim; the fifth city, Zoar, was spared due to some accident of its location. If some of the highly bituminous rock had been used in the construction of walls or buildings, it would have added fuel to the flames. Lightning may or may not have been partly responsible for starting the fires, but it would need only to have been seen in the sky at the time of the catastrophe to create the impression that the destruction came from the heavens.

The pillar of salt also has a geologic explanation, one which really belongs in the chapter on landform lore. Jebel Usdum, rising some 740 feet or so above the water level on the west side of the southern embayment of the Dead Sea, consists of rock salt overlain by gypsum-bearing marls (a marl is an impure limestone). Its best known feature is a pillar of salt, which has been separated from the main body of salt by erosion. It is not likely that this particular salt pillar has stood there for nearly four thousand years; not only does the region suffer frequent earthquakes, but salt is very easily eroded. A different pillar of salt on the south shore was described by Josephus about two thousand years ago and was still standing at least two hundred years later. In all probability there has been a succession of pillars of salt in the vicinity [34:900] throughout recent geologic history and all of human history, carved out of Jebel Usdum by winter rains. What could be more natural, then, than that such a noteworthy feature near the scene of a memorable catastrophe should become incorporated into the tradition of that catastrophe, or that it should be specifically associated with one of the most important victims, the king's wife?

So much for the folklore aspects of earthquakes. What do geologists say about their cause? Earthquakes occur when something happens in the earth to generate seismic waves; in most cases that something is *faulting*—the slipping of rocks past each other under stresses that have built up over a period of time. The place where an earthquake is generated in the earth is called the *focus* or *hypocenter* (Fig. 22), and it may be anywhere from very near the earth's surface to a few hundred miles deep. Earthquakes are classi-

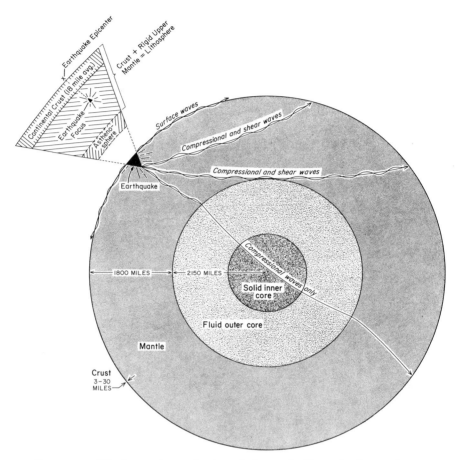

FIGURE 22. *The internal constitution of the earth. Most of what we know about the interior of the earth has been deduced from the behavior of the different kinds of seismic waves generated by earthquakes and, more recently, by large explosions—either nuclear or chemical (such as large quarry blasts or special seismic survey detonations).*

fied according to focal depth as *shallow* or *normal, intermediate,* and *deep-focus;* shallow shocks originate at depths down to 70 kilometers (about 44 miles), intermediate from 70 to 300 kilometers (44 to 188 miles), and deep at more than 300 kilometers (188 miles). The deepest yet recorded have been about 700 kilometers (440 miles) down. The place on the earth's surface directly above the focus is called the *epicenter.*

Magnitude is a quantity defining the amount of energy released at the focus. One of the largest magnitudes so far recorded is 8.6 or 8.7 for the Alaska earthquake of March 1964. Magnitude is most accurately determined from instrumental records, but it can also be estimated from the *intensity* and the distance to the focus. Intensity is a somewhat subjective measure of the force of an earthquake at the surface, estimated in terms of a scale based essentially on the extent of damage to man-made structures and the various effects felt by people in the area. In an uninhabited district, therefore, it is practically impossible to estimate earthquake intensity. Any given earthquake can have but one magnitude, but the intensity with which it is felt at any spot depends on the distance from the focus (intensity decreases in inverse proportion to the square of the distance from the focus; the epicenter, being closest, will be in the zone of maximum intensity), on the nature of the ground (vibrations will be of larger amplitude, and hence more destructive, in unconsolidated ground—especially wet ground—than in bedrock), and on the geologic structure of the region (seismic waves are less damped when traveling "with the grain" of the structure than against it). Usually the accumulated stress is not released all at once; the main shock may be preceded by a few smaller *foreshocks* and commonly is followed by a series of *aftershocks* of varying strength. Shallow earthquakes do their damage in a limited area around the epicenter; strong intermediate and deep-focus shocks, however, can have serious effects over extensive areas.

Earthquakes of the type we have been describing are called *tectonic*—because they have to do with deformation of the earth's structure—to distinguish them from those associated with volcanic activity or those few that are caused by landslides or the collapse of underground caverns. Earthquakes due to these last two causes are very weak and local and we can forget about them. Those preceding or accompanying eruptions are a special type. They are all relatively shallow, most of them extremely so, and therefore are never felt outside the immediate vicinity of the volcano. Rarely are they strong enough to cause damage. The question of volcanic versus tectonic earthquakes is rather important in connection with the Minoan eruption of Santorin, and will be given fuller attention in Chapter 8.

A corollary to Aristotle's explanation of earthquakes was that stifling, windless days precede a shock, all the winds having been driven underground. This concept lingers on in the widely held notion of "earthquake weather," as persistent a bit of modern folklore as can be found despite the fact that a check of the records undeniably shows that earthquakes have occurred at all times of the year, at all times of the day, and in all kinds of weather. C. McWilliams sums it up succinctly:

> In the popular sense, the phrase [earthquake weather] seems to designate a close, stifling, sunless, muggy atmosphere. One might be ready to believe that such weather does presage the coming of quakes if it were not for the fact that the description obviously refers to an atmospheric condition that, in California, is the subject of universal detestation. The conclusion is almost irresistible that the residents have merely made the loathsome "close days" responsible for calamities that could not, in loyalty, be imputed to any other kind of weather. [157]

There is only one way in which meteorological conditions might have some connection with earthquakes, and that is the possibility that barometric pressure might help trigger them, either independently or in conjunction with other forces such as tidal pull of the sun and moon. Not only the ocean but the solid earth too responds to the pull of the sun and moon, and just as ocean tides are higher at certain times depending on the relative positions of sun and moon, so are the tidal stresses acting on the solid earth stronger at certain times. Very slight extra stresses, due to these earth tides [177, 233] or atmospheric pressure or both, might in some cases be the last straw needed to release stresses that have been building to the breaking point in an earthquake region. The *trigger* of an earthquake, however, should not be confused with its cause any more than the trigger of a gun should be confused with the explosive that propels the bullet. Just as pulling the trigger of an empty gun will not cause it to fire, external forces acting on the earth cannot cause an earthquake if conditions within the earth are not ripe for one. Furthermore, the correlation between earth tides and earthquake occurrence is not a very strong one; numerous shocks have occurred when tidal forces were not at their maximum, or even when they were at their minimum. Thus the belief that the causes of earthquakes lie in the heavens is as unfounded as the belief in earthquake weather. At best, there is only the possibility, not yet proven, that planetary forces or meteorological conditions may help determine exactly when an earthquake will occur; but they have nothing to do with *why* they occur.

A correlation has been noted between earthquake frequency and the "polar wobble." The earth's axis is not perfectly fixed; it wobbles very slightly as it spins, as a top often does. Why it wobbles may have something to do with tidal frictions within the earth, or with movements of the liquid outer core. But whether the polar wobble causes earthquakes or is caused by them is currently a matter of debate. In any event, the ultimate *why* of earthquakes lies deep in the earth.

The leading misconception about earthquakes is the hideous notion that bottomless chasms can open in the solid earth, swallow up anybody or anything unfortunate enough to be in the wrong spot, and close up to crush the unhappy victim or object. Of all the very real dangers to be feared in an earthquake, however, that is the very least. It is true that cracks do sometimes open up in the ground in an earthquake, but never in solid rock. They are invariably superficial, and *they do not close up again* except under the rarest of circumstances. To be sure, someone observing a poorly constructed apartment building collapse within seconds to a heap of rubble, as has happened in countries without adequate codes of earthquake-resistant construction, might easily get the impression that the building had disappeared into the ground. But the fact is that sensational reports of people disappearing, or being caught up to their waists in the ground, and of houses or even whole villages being swallowed up and crushed, all date from before the nineteenth century and are all hearsay, usually at third or fourth or more remote hand. Authenticated reports are virtually nonexistent. Nevertheless many people in Japan, a country which has had more experience with earthquakes than almost any other, are convinced that they are safer in the house than in an open field, where they fear the ground might open beneath their feet.

L. Don Leet puts the mattter of earthquake cracks into proper perspective: "Sometimes the shaking . . . jars masses of unconsolidated materials so that they slump or slide into new positions. The natural cracks and fissures that result are not more fearful or re-markable than they would have been had the slump or slide oc-curred after a hard rain . . ." [139:44]. He describes one unusual case where cracks were observed to open and close again, in an abandoned rice field in Japan where the surface layer of dry loam was underlain by wet muck. During an earthquake, waves sloshed back and forth in the muck, causing the loam above to arch and crack and then close up when it settled back [139:45].

Only two reports of fatalities in earthquake cracks have ever been accepted by seismologists, and I have spent some time trying to track down the sources *behind* the published reports. One is the famous cow which fell into a crack during the 1906 San Francisco earthquake. Geologists are divided in their opinion as to the re-liability of this report. The official report on the earthquake con-tains the following statements:

> . . . at the Shafter ranch a fault crevice was momentarily so wide as to admit a cow, which fell in head first and was thus entombed. The closure which immediately followed left only the tail visible. At this point the fault-trace was a trench 6 or 8 feet wide, and the general level of the soil blocks within it was 1 or 2 feet below that of the adjacent undisturbed ground. [81:72]

Elsewhere in the same report it says:

> Mr. Payne J. Shafter's place is near the village of Olema. The fault trace is close to the house and other buildings. . . . During the earth-quake a cow fell into the fault-crack and the earth closed in on her, so that only the tail remained visible. At the time of my visit the tail had disappeared, being eaten by dogs, but there was abundant testimony to substantiate the statement. As the fault-trace in that neighborhood showed no cracks large enough to receive a cow, it would appear that during the production of the fault there was a temporary parting of the walls. [81:192]

The author of this part of the report, the eminent geologist G. K. Gilbert, received his information at second hand but evidently had no reason to doubt the word of his informants. However, in 1906 the idea of earthquake cracks that opened and closed again was not seriously questioned. Today such a report would be thor-oughly checked out by digging up the remains; there are those who believe Gilbert did just that, but his own field notebook contains nothing more than the statement "A cow was here swallowed by the crack, disappearing all but the tail . . . the testimony on this point is beyond question."

Even if the cow's remains had been dug up it need not have been proof that the cow was swallowed up during the earthquake. Robert Iacopi, author of the book *Earthquake Country* [120], received a letter from a Mr. H. H. Howard which casts serious doubt on the testimony of the witnesses. Mr. Howard, a cousin of the Shafters on whose ranch the alleged incident occurred, lived on the adjacent property as a boy. His letter* reads in part as follows:

> What I have wanted to write to you about is a memory of childhood touching on your text page 147; a memory which came flooding back to me each year when, as I lived in the middle west, people would send me the annual Earthquake and Fire editions of the San Francisco newspapers wherein would be faithfully chronicled that lovely tale about how the earth swallowed the cow! . . . One lovely warm day, I can remember this quite plainly though I was very young for this could not have been later than about 1912 or 1913, my father and I were sitting on a bench in the garden . . . when our cousin Payne Shafter rode up. . . . The two men talked briefly, and then for no reason I can remember my father said to Payne, I paraphrase, "Payne, why on earth did you tell those reporters that time that your cow was swallowed up by that crack in the earth?" To which Payne replied, I again paraphrase, "Look, Pax, the cow had died, and we had to bury her. That night along came the earthquake which opened up a big crack in the ground; we simply dragged the carcass over to the crack and tipped it in, with the feet sticking out. Then along came those newspaper reporters and when they got the idea that the cow had fallen in we weren't about to spoil a good story. Why spoil it now?"

> I haven't the foggiest first hand knowledge as to what actually did happen, but I know for certain that the conversation described took place because I heard it and heard my father laugh about it a number of times afterward. . . . I admire your caution in reporting the incident in your book and leaving the "official" story in quotes!

* This letter to Mr. Iacopi ended up in the files of the U.S. Geological Survey's Office of Earthquake Research in Menlo Park, California, and was called to my attention by a colleague there. In giving permission to quote from it, Mr. Howard remarked that the letter certainly got around, for he received numerous inquiries about it from all up and down California. He emphasized that he can only report what he heard the two men say, but added that he can think of no reason why what they said was anything but the truth.

This letter at least raises the question as to whether Gilbert was shown the same crack into which the cow had disappeared, dead or alive, and leaves us wishing that he had seen that tail sticking out of the ground with his own eyes.

The only other documented case, and the only one involving a human fatality, is that of a woman killed in the Fukui earthquake of June 28, 1948, in Japan. There is no doubt about what happened, but even in the official report on the Fukui earthquake [248] there is some conflict of opinion as to how it happened. The report speaks for itself (slightly paraphrased in the interests of correct English):

There is universal fear of being engulfed in ground fissures which appear during earthquakes. But we have no such record in the earthquake history of this country. Imamura, therefore, affirmed that there is no need of this fear, although he himself had reported a case of smaller fissuring which took place in the Kwanto earthquake of 1923 in the school yard at Hojo in Chiba prefecture, in which two lines of fissures opened and closed spouting water intermittently. But a tragic case actually occurred in the present Fukui earthquake. A peasant wife, 37 years of age, was working in the rice paddy at 33 Wada-shussaku-machi, in Fukui City, when the shocks were felt. She was found dead having been buried up to her chin in a fissure of about 100 m. length. It is believed that she was crushed to death but not to have been drowned. It is said that a fissure of about 4 feet in width opened and then closed. After the earthquake, only a faint reminder (about 2 cm. in width) of the course of the fissure could be traced at the spot. . . .

It [the opening and closing of fissures] may be attributed to the gravity waves set up in the soft ground . . . the cracks at some places were arranged in conformity with the old river course even though it is filled at present. . . . This may be considered as evidence of the relative motion of the soft fill due to the violent earthquake motions. If the motions were repeated more than once, the soft fill would oscillate back and forth so that cracks made therein might also open and close successively. [248:26-27]

The above words appear in the general description of the earthquake compiled by H. Kawasumi. In a later section of the same report, dealing specifically with the cracks and fissures opened up during the earthquake, N. Miyabe prefers a different explanation of the woman's death:

The neighborhood of Fukui City is a low alluvial swampy plain. During the earthquake, ground water gushed out at so many places there and some of them formed mud volcanoes. The bearing power of the ground was [reduced] by the water, so some electric cars near Fukui under which the water spouted sank along their wheels. . . . A woman was found dead almost sunk in the earth near Fukui after the earthquake. The people said she was wedged by the closing of an earthquake fissure. But the writer considers that she would have sunk in the soil during the earthquake when the bearing power of the soil became very small [due to] water spouted as mentioned above, taking account of the following fact: even in usual times some swampy rice fields near Fukui are so soft that men are in danger of sinking in the soil unless they walk on long bamboo or wooden rods laid upon the ground. [248:156-157]

Of the two explanations, the latter seems more logical. If a crack had yawned at the woman's feet, would she not have been thrown off balance and pitched headlong into it, or across it, or alongside it, flinging out her arms as she fell? But she was found in a vertical position, which suggests that while she was upright, probably trying to walk or run out of the rice field, she was literally shaken down into the soft earth, as though in a quicksand, and crushed by the weight of the thick mud pressing in around her body. Some mechanism of this sort might also be the explanation for old reports of people being buried up to their waists in the ground, if indeed there is any truth to them at all.

Large cracks opened in the ground during the New Madrid earthquakes, and remained open; some can still be seen today. Not all of them were simple fissures; a number were down-dropped fault trenches or "grabens." The deepest were not more than twenty feet deep [65:52]. Although there were many wild rumors of people having been lost in these fissures, all those who managed to fall into them were extricated, though sometimes with difficulty. All but two of the casualties reported in the New Madrid earthquakes were drownings, caused when river bluffs caved in or when boats or islands in the river were swamped. (One person was reported killed by a falling wall, and one woman was so frightened that she ran until she collapsed and died of fear and exhaustion [65:43]). Still, the fear of being entombed in the earth was uppermost in everyone's mind, and observing that the "chasms" ran consistently in one direction, people felled trees at right angles to that direction, and when a shock began they rushed out and stationed themselves on those tree trunks [65:47]. Many believed that they saved their lives thereby. One thing that did disappear into the ground in one of the New Madrid shocks was a load of heavy castings in a cellar [65:52]; these, like the woman in Fukui, could have been shaken down into the unconsolidated ground during the vibration. Several small islands in the Mississippi River disappeared during the earthquakes, and popular accounts leave the impression that they were gulped down into the earth. Some may have been down-dropped below water level, but what could have happened in many cases is that, being made of soft, unconsolidated, saturated ground—the most unstable there is in an earthquake—the islets were literally shaken to bits and washed away piecemeal.

The most likely way in which an earthquake crack could cause a fatality these days would be if a car should run into one which suddenly opened in a highway. Cracks in paved roads are one of the most common forms of earthquake damage, and many are big enough to throw a car out of control or turn it over. In the highly unlikely event that a crack did open beneath one's feet, due to slumping of unconsolidated material, the most probable injury would be a nasty bump or at most a broken bone or two. The point to be remembered is that if cracks open in the ground during an earthquake, they are almost certainly going to stay open (Plate 19). Your chances of being swallowed up in an earthquake crack are infinitesimally smaller than, for instance, your chances of being struck by a falling aircraft.

The very real dangers in an earthquake are: (1) being struck by falling objects, particularly as you rush outdoors; (2) being burned in a fire resulting from the earthquake, a danger which frequently is compounded by the disruption of water mains; (3) being buried in an earthquake-triggered landslide if you are in a mountain area, like the nineteen people sleeping in a campground on Hebgen Lake in Montana on the night of August 17, 1959 [94], or the tens of thousands of inhabitants of Yungay and Ranrahirca in Peru who were buried by the Huascarán debris avalanche triggered by the earthquake of May 31, 1970; and (4) being drowned in a tsunami, if you are on the coast. Therefore, if you should happen to be out in the open on a plain far from the sea when an earthquake occurs,

PLATE 19. *One of the larger fissures opened during the Alaskan earth-quake of March 1964. This fissure, 3 feet wide and 7 feet deep, is in un-consolidated glacial till along the trace of the Patton Bay Fault, Montague Island. Note that it is neither bottomless, nor has it closed up on the trees toppling into it. (Photo by George Plafker, U.S. Geological Survey.)*

you can just relax and enjoy it! And for final reassurance about earthquake cracks, just consider this: in the 1954 Dixie Valley earthquake in Nevada, a small sturdily built wooden cabin stood on a segment of ground which dropped down to form a small graben. Although the cabin was only a few feet from the twelve-and-a-half-foot fault scarp that was suddenly created, not a windowpane was cracked, and a china cup did not even fall off its shelf [226].

In contrast to the idea of the yawning earth, which certainly is largely—and in my opinion, entirely—folklore, there are examples of earthquake folklore which may yet turn out to be factlore. Linked to the belief in the *namazu* as the causer of earthquakes in Japan, there have been widespread reports of unusual behavior of catfish preceding major earthquakes there. It was reported, for instance, that just before the earthquake of 1855 a fisherman was struck by the agitation of the catfish and, believing that it portended an earthquake, hurried home just in time to save his family and possessions [173:56]. Other reports have told of marked restlessness of catfish the day before the terrible Tokyo earthquake of 1923. The belief that catfish can somehow sense an earthquake before it happens has been so persistent that scientists began to wonder whether there might be some germ of truth behind the idea. Accordingly, nearly forty years ago Professor Hatai [100, 101, 102] and his colleagues at Tohoku University began systematic observations of tankfuls of catfish on a laboratory table. They found that normally the fish remained perfectly quiet and did not react to a gentle external stimulus in the form of a finger tap on the table; however, they occasionally gave a perceptible twitch or jump when the table was tapped, and 80 percent of the time this sensitivity preceded an earthquake by six to eight hours. How the catfish knew that an earthquake was about to occur may have had something to do with natural electrical currents in the earth, for when the water circulating through the tanks was not passed through the ground, the sensitivity virtually disappeared.

If this ability of the catfish to sense impending earthquakes is real, there need be nothing supernatural about it. In recent years seismologists interested in earthquake prediction have been investigating, among other things, the *seismoelectric* [205] and *seismomagnetic* [204] effects. The stresses accumulating in rocks, which eventually are released in the form of an earthquake, produce slight changes in the earth's magnetic field or in its electrical conductivity, which can be monitored by sensitive instruments. Not only catfish but other water-dwelling creatures as well may possess nerve cells capable of perceiving these or other natural phenomena which humans can not. (In the catfish these nerve cells are believed to be in the feelers and along the sides of the body.)

Along these same lines an American, B. H. Armstrong [3], has investigated the possibility that reports of unusual agitation in animals before earthquakes might have a physical basis. There are stories of how dogs in Talcahuano, Chile, ran howling before the Concepción earthquake of 1835, and of how dogs, cats, and cows awakened people in Taal, in the Philippines, shortly before an earthquake there. Armstrong was interested in the "seismoacoustic" effect, which is the release of high-frequency sound waves, beyond the human auditory range, resulting from strain buildup in rocks. Acoustic emission from strained rocks is already monitored by instruments in mines, to warn of impending rockbursts.* Although the results obtained in Armstrong's experiments were not conclusive, the possibility could not be ruled out that for earthquakes involving surface rupture, or earthquakes with rupture occurring in a body of water (bringing us to fish behavior again), the buildup of stress leading to the earthquake could cause a change in the acoustic absorption properties of rock sufficiently substantial to cause the emission of sound waves perceptible to animals.

* Rockburst is the sudden and often violent failure of rocks in mines or quarries, which can be fatal to miners and which on a small scale resembles a natural earthquake.

6 /

Volcano Lore

$\mathbf{P}$RIMITIVE PEOPLES LIVING IN VOLCANIC AREAS, subjected repeatedly to incredibly awesome and frequently dangerous displays of forces in the earth whose true nature they could not possibly comprehend, quite naturally tended to attribute those manifestations to the activity of deities or demons who were malign or, at best, merely capricious. The Aztecs, Mayas, and Quechuas ("Incas") offered human sacrifices to volcanoes [188], and until recently so did the peoples of many other active volcanic areas. The Javanese sacrificed human beings to Mount Bromo, and they still throw live chickens into the crater once a year [108:25]. In the Congo, the cruel god Nyudadagora every year claimed as his slaves ten of the finest young men of the tribes living near Nyamuragira and Nyiragongo, north of Lake Kivu (see Fig. 25, this chapter); after savage and brutal rites they hurled themselves into the abode of their new master [108:24]. In Nicaragua it was formerly believed that Coseguina would stay peaceful only if a baby was thrown into the crater every twenty-five years [108:31]. If in any of these places the sacrifices failed to forestall or halt an eruption, it could always be claimed that things might have been even worse without them.

Myths and legends about volcanoes are of several types. Some try to account for the existence of individual volcanoes or volcanic landforms, and thus are clearly etiological. Some seek a reason for volcanic activity in general, or for individual eruptions in particular, and these also are etiological. But more than a few may be reports of real eruptions, in which the historical basis has become unrecognizable. All these types of geomyth may occur in the same region, as some of the examples below will illustrate.

The volcano deity best known to us is the goddess whose lore so permeates our fiftieth state that even the most casual visitor to the Hawaiian Islands will have heard of her; thus it should not be out of order to devote a disproportionately large amount of space to Pele. To this day she is feared and revered by many Hawaiians. In August 1881, when lava from Mauna Loa threatened the city of Hilo, an appeal was sent to Princess Ruth Keelikoani, one of the last of the Kamehameha royal line. The sixty-three-year-old princess hastened to Hilo and approached the edge of the advancing lava flow, where she chanted ancient incantations, offered silk scarves to Pele, and finally emptied a bottle of brandy onto the creeping destruction (just as her ancestors had once poured libations of *awa*, an intoxicating drink made from the roots of *Piper methysticum*, a species of pepper); next morning the lava stopped short of the city [108:35]. Even as recently as 1955, when a lava flow menaced the village of Kapoho, the inhabitants chanted at the edge of the lava stream and offered food and tobacco to Pele; that time also the lava stopped short of the village [108:35].

There are innumerable variants of the story of the coming of Pele to the Islands, and nobody knows precisely whence she came, except that it was from somewhere farther to the south—some say Tahiti. Apparently a bitter quarrel with her older sister Namakaokahai forced her to seek a new abode. Pele carried a magic digging tool called Paoa, and wherever she attacked the earth a volcanic crater opened up. She tried first to settle on the island of Kauai (Fig. 23), and there dug a deep pit. The earth thrown out of that pit formed the hill known as *Puu-ka-Pele* ("Pele's Hill"). (Puu-ka-Pele is the eroded core of a volcanic cone from which issued part of the lavas of the Waimea Canyon series, the rocks which have built the principal volcanic mountains of Kauai.) Namakaokahai pursued and set upon Pele and left her for dead, but Pele recovered and moved on to Oahu. There she dug a fire pit at Moanalua, near Honolulu, but that crater filled with salt water and became Keealipaakai, the Salt Lake; the material dug out of it created the hill called the White Bird. (Keealipaaki and Keealiamanu

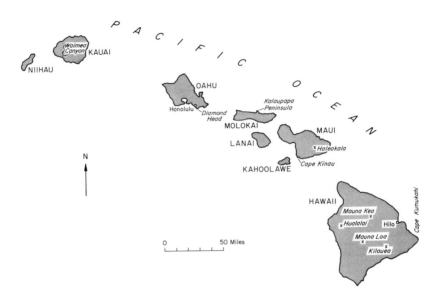

FIGURE 23. *Sketch map of the Hawaiian Islands archipelago. Legends concerning the coming of Pele to the islands show that the Hawaiians were fully aware of the fact that volcanic activity has migrated southeastward along the chain of islands. At present only Mauna Loa and Kilauea are active, although Hualalai and Haleakala each have had one eruption in historical times and thus are classed as active.*

are two overlapping craters which, together with a third called Makalapa, constitute the Salt Lake Craters, the westernmost center of recent pyroclastic* activity on Oahu. The White Bird is the highest point on the rim of Keealiamanu.) No matter how deep Pele dug in the mountains she found no fire, but along the coast she had better luck. At Leahi, better known now as Diamond Head, large amounts of material were erupted from her pit until she struck water, which quenched the fires.

Diamond Head is an extinct crater formed about a hundred and fifty thousand years ago in what is known as a *phreatomagmatic* eruption—one in which water comes in contact with hot rising

---

* Pyroclastic eruptions are those in which only solid material is ejected. As a rule they are very much more explosive than lava eruptions. The word *pyroclastic* is also applied to the products of such eruptions.

magma.* Sea water and ground water poured down the crack along which the magma arose, resulting in violent steam explosions. The cone, built up of soft layers of unconsolidated volcanic ash, lapilli,† and fine shattered fragments of the reef limestone penetrated by the magma, was quickly eroded by running water and sea waves. All that is left of it now is the crater itself and its immediate rim. The name Diamond Head was given to this prominent landmark by British sailors, because calcite crystals derived from the limestone sparkled like diamonds in its walls.

So Pele passed on to the next island, Molokai. There she dug Kauhako crater (whose flows have built the Kalaupapa peninsula on which lies the famous Leper Settlement), but again she struck water and was disappointed. Next she tried Maui, into the top of whose mountain she dug to create the great volcano Haleakala. When Namakaokahai saw the smoke rising from Haleakala she realized that her sister was still alive. Again they fought, and again Pele was destroyed. Her bones were scattered along the coast, forming the lava islets known as *Kaiwi o Pele* ("Pele's Bones"). Namakaohakai retired rejoicing in victory, but when she looked back over the sea she saw Pele's unconquered spirit form hovering in clouds of smoke and flame over Mauna Loa on the island of Hawaii (Plate 20). Then she realized that she could never overcome the

---

* Magma is molten rock materal, generated deep in the earth when pressure and temperature conditions are favorable for melting. Magma may solidify into rock deep below the earth's surface, in which case it forms coarse-grained rocks, or it may reach part way or all the way to the surface. Lava is magma which reaches the surface in fluid form.

† The fragmental materials ejected from volcanoes—that is, everything other than fluid lava and gases—may be of all sizes. *Blocks* are masses of pre-existing rock torn from the substratum or from the walls of the vent; they may be anything up to the size of a house. *Bombs* are masses of molten lava blown from the vent and solidified during their flight through the air or after landing. *Lapilli* are bombs ranging in size approximately from that of a pea to that of a walnut. Anything finer than that is called *ash* (sometimes the coarser ash particles are called *cinders*). The finest ash particles can remain suspended in the atmosphere for a very long time and may be carried hundreds of miles by the wind. All the fragmental ejecta are collectively termed *tephra* (the Greek word meaning ash). The term *pyroclastic* (defined above) includes not only fresh-fallen tephra but also the ejected material after it has been compacted into rock, either where it fell or after being eroded and redeposited or otherwise "reworked." *Tuff* is the name given to the soft rock made essentially of compacted volcanic ash.

PLATE 20. *Pele, the Hawaiian volcano goddess, as imagined by D. How-ard Hitchcock. The artist explains why he depicted her as white in these words: ". . . that's exactly as I saw her. Her skin was light, because it was fire. Her hair was swirls of smoke and flame, and fire dripped from her fingers." Hawaiian tradition actually does tell that Pele was a fair-skinned foreigner, and some ancient poetry describes her as a tall blond woman without eyebrows. (Reproduced by courtesy of the U.S. National Park Service.)*

fire goddess, and left her in peace thenceforward. Pele dug her final fire pit, Halemaumau, in the floor of the Kilauea caldera and brought the rest of her numerous relatives to live with her. There they still dwell [270:4-13].

This legend clearly shows that the Hawaiians were very much aware of the fact that the volcanic activity of the islands is progressively younger from northwest to southeast (See Fig. 23). An intelligent "sophisticated savage" would not have needed formal geologic training to deduce that fact, from the relative degree of erosion of the different volcanoes, from the extent to which they and their flows are overgrown with vegetation, and from the relative freshness of their lavas. Only four volcanoes are active, three on the Big Island (Kilauea, Mauna Loa, and Hualalai) and Haleakala on Maui. Mauna Loa and Kilauea account for all the activity at present; Hualalai and Haleakala each have erupted only once in historical time, Hualalai in about 1800 and Haleakala some time between 1786 and 1793.

Kilauea has been far more spectacularly active than Mauna Loa, and its folklore reflects its geologic history rather faithfully. In the fourth volume of his *Polynesian Researches*, published in 1833, William Ellis [56], a British missionary who understood the Hawaiian language and first recorded many of the Hawaiian legends and traditions, vividly described the fiery display in Halemaumau and then went on to say:

> The natives . . . sat most of the night talking about the achievements of Pele, and regarding with a superstitious fear, at which we were not surprised, the brilliant exhibition. They considered it the primeval abode of their volcanic deities. The conical craters [not actually craters, but mounds built up by lava fountains], they said, were their houses [Plate 21], where they frequently amused themselves by playing at konane [a game resembling checkers]; the roaring of the furnaces and the crackling of the flames were . . . the music of their dance, and the red flaming surge was the surf wherein they played, sportively swimming on the rolling wave [Plate 22]. . . .
>
> We learned that it [the volcano] had been burning from time immemorial . . . and had overflowed some part of the country during the reign of every king that had governed Hawaii; that in earlier ages it used to boil up, overflow its banks, and inundate the adjacent country; but that for many kings' reigns past it had kept below the level of the surrounding plain, continually extending its surface and increasing its depth, and occasionally throwing up, with violent explosions, huge rocks, or red-hot stones. . . .
>
> No great explosion, they added, had taken place since the days of Keoua [see Chapter 2]; but many places near the sea had since been overflowed, on which occasions they supposed Pele went by a road under ground from her house in the crater to the shore [56:182-183].

What apter description of a flank eruption* could be found, in a

---

* An eruption in which lava bursts out through a vent or fissure in the side of the volcanic edifice. Flank eruptions have been quite common in recent years along the East Rift extending from Kilauea to Cape Kumukahi (Fig. 24); the longest known eruption of Kilauea, the Mauna Ulu eruption which began in May 1969 and is still in progress as this is written, is on this rift.

PLATE 21. *Halemaumau, the "fire pit" of Kilauea volcano, during an eruption. The conical mounds, which Hawaiians believed were the dwellings of Pele and her family, are transient features built up by lava fountaining. (Photo by Donald A. Swanson, June 1968. Reproduced by courtesy of the Hawaiian Volcano Observatory, U.S. Geological Survey.)*

PLATE 22. *Lava lake in Halemaumau, Kilauea volcano. Until 1924 molten lava boiled constantly in Halemaumau, but since then the spectacle first described by William Ellis can be seen only during periods of active eruption. (Photo by Richard S. Fiske, November 1967. Reproduced by courtesy of the Hawaiian Volcano Observatory, U.S. Geological Survey.)*

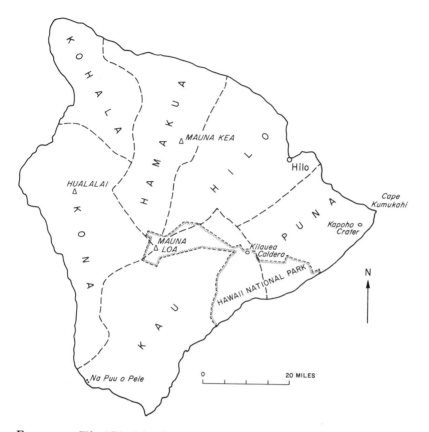

FIGURE 24. *The "Big Island" of the Hawaiian Islands archipelago, showing the location of places associated with various volcano legends.*

figurative sense, than "Pele went by a road under the ground from her house in the crater to the shore"?

Pele's fire pit Halemaumau was almost continuously active from the time the first Europeans visited Kilauea, in 1823, until 1924. Repeatedly, lava welled up and filled the pit, sometimes overflowing and flooding the entire caldera floor; repeatedly the pit was reestablished as its floor sank into the drained space below. Since 1924, however, the surface of the lava lake has been crusted over with a thick layer of dark gray congealed lava. Spectacles of the kind witnessed by Ellis are seen only during specific eruptions, when the hard gray crust is riven by fissures along which bright red incandescent lava wells up, or brilliant fire fountains leap up to heights of hundreds of feet. Chilled upon contact with the air, the bright surface of the lava lake quickly films over with gray, but this new-formed crust cracks as the molten mass beneath it boils and surges; slabs of crust float for a time on the glowing liquid, tilt, and slowly slide beneath the surface. It is a spectacle never to be forgotten, even if seen only on film without the accompanying roaring sounds and choking fumes.

Pele is a very temperamental deity, easily enraged to the point where she sends floods of lava to destroy the object of her displeasure, often destroying scores of innocent bystanders besides. When she stamps her foot in anger, as she often does in calling up an eruption, the ground trembles. (Volcanic eruptions as a rule are preceded and accompanied by local mild earthquakes.) According to one legend the very desolate Kahuku area on the southwest coast of Hawaii was once lush and green, with many trees and flowers and with fields of sugar cane and taro which supported several villages. Pele destroyed it all in a fit of pique [270:23-25]. She had become enamored of two young chiefs of the region who excelled at all sports including *holua* racing.* Appearing as a beautiful princess, Pele joined the chiefs and their companions in that sport, in which she easily surpassed all the other women.

Pleasure-filled days went by, but the young men began to suspect the identity of the stranger, whose moods changed so fitfully, and sought to avoid her. As their reluctance to race with her increased, so did her annoyance, until the ground heated up and the grass died. At this the chiefs took fright and tried to run away, but Pele stamped until the ground shook and a flood of lava burst forth and overwhelmed all of Kahuku. As the chiefs sought to escape into the sea, first one and then the other was locked in Pele's fiery embrace, then tossed aside, and the lava piled up around their bodies. And there they still stand, as *Na Puu o Pele* ("The Hills of

---

* A *holua* was a long, narrow wooden sled used for coasting down grass-covered slopes or specially paved hillside runways (Plate 23). The rider took a running start and fell flat on the sled, just as children go "belly-whopping" on snow sleds. As *holua* sleds were only two to four inches wide in front and about six inches wide at the rear end, and anywhere from seven to fourteen feet long, it required considerable skill to maintain balance on one. An expert *holua* rider could coast as much as two hundred yards.

PLATE 23. Holua *slide at Puu Hinahina, Kapu'a, S. Kona, Hawaii. At bottom: a* holua *sled, used to coast down grass-covered slopes or runways like that shown above. In Hawaiian folklore, Pele often competed in* holua *races, and caused eruptions if displeased with the outcome. (Photos by courtesy of the Bernice P. Bishop Museum, Honolulu.)*

Pele"), two symmetrical mounds on the shore, surrounded by a sea of barren *aa* lava.

As a matter of geologic fact, Na Puu a Pele are the largest of a series of littoral cones along the Kahuku coast. When hot lava enters the sea, particularly an *aa* flow, whose fragmental surface allows sea water to penetrate readily to the hot center of the flow, steam explosions occur. Littoral cones are formed if the fragments of lava comminuted by the explosions pile up on the surface of the flow. The Kahuku area is truly one of the most desolate parts of Hawaii, covered with bleak lava flows the latest of which was erupted in 1907. Even the prehistoric flows are relatively barren of vegetation, for by virtue of its location with respect to the great mountain mass of Mauna Loa and the prevailing winds, the Kahuku area receives little rain, and what does fall sinks deep into the pervious lava. The legend of Na Puu o Pele is not too consistent with the geologic facts on two counts: the Kahuku area could never have been lush and green within the memory of the Hawaiians, and the hills themselves are not made of lava, but are heaps of fragmental material.

Cape Kumukahi owes its name to another handsome chief who took Pele's fancy and suffered a similar fate [207:27-29]. Having known her only in the guise of a young and beautiful princess, Kumukahi made the fatal mistake of ridiculing Pele when she appeared as an old woman who demanded to be allowed to participate in sports with him and his friends. Instantly a fire fountain burst forth; Kumukahi headed for the sea, but Pele caught him on the beach and heaped a mound of lava over him, while a flood of lava poured around him and created the cape. (The cape really has been built outward by repeated flows from the East Rift of Kilauea; as recently as 1960, about five hundred acres of land were added to its northern side.)

Yet another chief, Papalauahi by name, incurred Pele's displeasure when he outshone her at *holua* riding. This time the lava caught some of the other chiefs who were racing with him, and turned a group of frightened onlookers into pillars of stone [270: 29-30]. The pillars of lava which figure in this legend are tree molds (Plate 24). Lava tree molds are hollow cylinders, perhaps up to two feet in diameter, with walls one to six inches thick, standing as much as ten to fifteen feet above a lava floor. When lava flows through a forest, it becomes chilled upon coming in contact with a tree trunk and solidifies to form a crust around the tree. When the lava drains away, that crust remains standing upright. If it drains away rapidly, as it would on a steep slope, charred trees may be preserved within the mold; but if the lava forms a lake and stands for a time, the trees burn away completely, leaving only the impression of carbonized wood on the inner surface of the mold [165].

Inasmuch as Pele was known to have such a taste for handsome young chiefs, you would think that any such would have learned to

PLATE 24. *Tree molds in Lava Tree State Park, Puna district, Hawaii. These were formed during an eruption in 1790 and may be the "petrified people" who figure in the legend of Papalauahi. (Photo by C. J. Vitaliano, August 1970.)*

be polite to strange females, young or old. But apparently not. Another who ran afoul of her wrath was Kahawali, a chief in the Puna district who was an expert *holua* rider [270:37ff.]. With his best friend, Ahua, he spent many hours racing down the slope of a hill near Kapoho. One day a woman of undistinguished appearance asked if she might try his sled. Disdainfully he refused; a chief's property was not for the use of any common person! Flinging himself down on his *holua*, he sped off down the slide. At this the woman's eyes flashed in anger and she stamped on the ground. The hillside was rent by an earthquake, lava gushed forth and poured down the *holua* course, and Pele in her terrible supernatural form rode her sled down the wave of fire. Calling to Ahua, Kahawali fled toward the sea. At the small hill known as Puukea he threw off his cloak. At Kukii he touched noses with his mother, then sped on to his own house. His wife implored him, "Stay and let us die together," but he ran on, passing his sister on the way. Then he met his pet pig, Aloipuaa, with whom he touched noses, and ran on; minutes later the pig was engulfed and turned into a big black stone. At the shore he commandeered a canoe which his younger brother had procured to save his own family, and Kahawali and Ahua paddled rapidly to the sea. Pele hurled huge rocks after their retreating forms, but none scored a hit. It is said that these rocks can still be seen under water.

The crater pointed out to Ellis, who first recorded this legend, as the scene of the *holua* race was described by him as "a black and frowning crater, about one hundred feet high, with a deep gap in its rim on the eastern side, from which the course of the current of lava could be distinctly traced" [56:223]—in other words, a typical example of a cone breached by a lava flow. The hill designated *Kaholua o Kahawali* ("Kahawali's Sliding Place") on modern topographic maps does not fit that description, although it is in the right position with respect to the other places named. The Kaholua o Kahawali of today is just the remnant of an old cinder cone, much of which has been removed to build roads through the sugar cane fields which surround it. Tropical vegetation crowns what may have been its original summit, while its bare gouged-out flanks are weathered to a reddish brown. Any evidence of a lava flow issuing from it has long since disappeared under the cane fields (for productive soil develops rapidly on this well watered side of the island), or under the 1955 and 1960 lava flows which now cover much of the scene of Kahawali's activities. Aloipuaa's rock, in the center of a lava channel, and other rocks scattered along the banks of that channel which were once pointed out as people and houses destroyed by Pele [270:43], are also buried under those flows. Puukea is still there, however; like Kahawali's Sliding Place, this cone wears a heavy coat of vegetation except on one side where its cinders have been quarried to build roads. Puukea is a younger (but still prehistoric) cone on the flank of the older and larger Kapoho crater, and may date from about the same time as the eruption which engendered the Kahawali legend; in that case it could have been the source of the rocks (presumably volcanic bombs) which Pele is supposed to have hurled after the fleeing offender. Kukii is the site of a *heiau* or shrine on another small overgrown cone, Puu Kukau, now surrounded by the 1960 lava flow which extended Cape Kumakahi seaward.

Ellis surely was correct when he inferred that this legend probably was based on some "sudden and unexpected eruption of a volcano, while a chief and his people were playing at *holua*" [56:222], for several flank eruptions have issued in this area. As mentioned at the beginning of the chapter, lava from the 1955 eruption menaced but did not quite reach the village of Kapoho; but the 1960 eruption completely obliterated it. In 1960 the destruction of the village was news; similar events in prehistoric time, however, gave us the legends of Kahawali, Papalauahi, and Kumukahi, all of which pertain to the area along the Kilauea East Rift. As a missionary, Ellis was dismayed by what the story of Kahawali revealed about the character of the Hawaiians. "The absence of relative affections shown by Kahavari [Kahawali], who, notwithstanding the entreaties of his wife, could leave her, his children, his mother, and his sister to certain destruction, meets with no reprehension; neither is any censure passed on his unjust seizure of the canoe belonging to his brother, who was engaged in saving his own family, while his adroitness at escaping the dreadful calamity of which he had been the sole cause is applauded in terms too indelicate to be recorded" [56:222-223]. This same ethic, so decidedly unlike the Judeo-Christian, is also apparent in Polynesian flood traditions, as shown in Chapter 7.

For a while Pele was married to the demigod Kamapuaa, who was a *kupua,* a being who could appear either as animal or man; usually he was an attractive human being, but in his brutish form he assumed the shape of a hog. During this interlude Hawaii was divided between Pele and Kamapuaa, she taking the Puna, Kau, and Kona districts (see Fig. 24), where lava rocks are most plentiful (these are the districts where the most recent volcanic activity has occurred), and he the Kohala, Hamakua, and Hilo districts, which are free of fresh lavas. The marriage between these two strong-minded individuals was stormy and short-lived. Often they fought. In mighty battles of the elements, earthquakes and fire were met by floods from the sea and from the skies (over which Kamapuaa had some influence). Eventually Kamapuaa was routed from the Pit of

Pele, pursued by streams of lava. The lava extended to the sea on either side of him, cutting off retreat by land and heating the water to boiling to preclude his escape by sea. Kamapuaa changed himself into a fish with a very thick skin and thus was able to swim through the overheated waters to safety—and that is how the Hawaiians account for the fish known as the *humuhumu-nukunuku-a-puaa*, an angular, thick-skinned fish which emits a grunting noise (hence its name, which means "grunting angular pig"). Right down to the present time, a hog has always been regarded as the offering most acceptable to Pele [270:45-54].

Every new Hawaiian eruption spawns new tales of Pele. An amusing story is recounted by Dr. Gordon Macdonald, volcanologist at the University of Hawaii. In his words, it

tells of a man from Kohala, at the north end of the island of Hawaii, who was sitting near the edge of the lava flow of 1926, watching the moving flow, and nipping at a bottle of (then illegal) gin. One of his friends said to him, "Don't you know it's unlucky to drink that without giving some to Pele?" That worried him, so after taking another swig, he corked up the bottle, held it up and noted to his regret that it was still half full, and threw it into the lava river, calling out, "This is for you, Pele!" It disappeared, but after perhaps 30 seconds it came sailing out again and landed in the grass at his feet. He picked it up, and observed that it was now less than a quarter full. Unfortunately, there is no record of what happened to the rest of the gin!

Pele has lent her name to two volcanological details observed in connection with basaltic lavas. Gas bubbles bursting in an active vent or at the surface of a boiling lava lake or stream throw out droplets of the extremely fluid melt, and these droplets may be borne upward on strong thermal updrafts (as in a flue) and chilled into shining, greenish-black or brownish-black glassy globules known as *Pele's tears*. More often than not, a tear trails a thread of spun glass. These glassy threads break off very easily and drift together to form masses of *Pele's hair*. Pele's hair can also form from fountaining lava, in much the same way that the insulating material known as rock wool is manufactured. The latter is made by blowing a jet of steam into molten rock, which is essentially what nature does in a fire fountain. A strand of Pele's hair is the tiniest of volcanic bombs. The terms *Pele's hair* and *Pele's tears* have been adopted into the technical vocabulary of volcanology. Plate 25 shows Pele's hair and Pele's tears from Kilauea, some still connected.

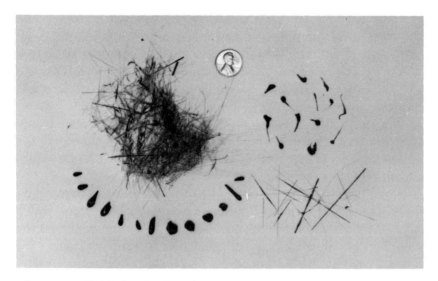

PLATE 25. *"Pele's hair" and "Pele's tears," formed from wind-swept molten lava. (Photo by C. J. Vitaliano.)*

New Zealand boasts of several active volcanoes and one of the world's great thermal areas, the closest rival to Yellowstone National Park in such features as hot springs, geysers, boiling lakes and pools, fumaroles, and other manifestations of underground heat connected with the dying stages of volcanism. The Maoris, who were living there when the British came to explore and colonize, are a Polynesian people like the Hawaiians. Inevitably, Maori folklore is rich in legends concerning the volcanoes. One of the best known of their tales is a variation on the theme of the eternal triangle [49:86]. The volcanoes Taranaki, Ruapehu, and Tongariro (later called Ngauruhoe; see Fig. 2) were giants who once all lived in the same area. Taranaki and Ruapehu both fell in love with Tongariro, but she could not decide which of them she preferred. At length they decided to fight for her. Taranaki tore himself loose from the earth and launched himself at Ruapehu, trying to crush him; but Ruapehu countered by heating up the waters in his crater lake and spraying scalding water over Taranaki and the surrounding countryside. Taranaki, enraged and in pain, hurled a shower of stones which broke the top of Ruapehu's cone, ruining his good looks. Ruapehu swallowed the broken cone, melted it, and spat it at Taranaki, who was forced to repair to the sea to ease his burns. His path to the sea is the Wanganui River valley. He retreated up the coast to his present location in the province which bears his name, where he stands brooding on revenge. Superstitious Maoris will not dwell or be buried anywhere on a line between Taranaki and the other two peaks, for some day, they believe, he might return.

Taranaki, now known as Mount Egmont, rises majestically from sea level to an altitude of more than eight thousand feet, a snow-capped, symmetrical cone often compared to Japan's Fuji-yama. Winds sweeping in from the Tasman Sea are forced to rise over the peak and, becoming chilled at high altitudes, drop their moisture to make Taranaki Province a rich green pastureland which supports much of New Zealand's dairy industry. So well does the mountain act in its capacity of rainmaker that its handsome head is usually hidden in the clouds. There is no historical record of any eruption of Mount Egmont, but one of its freshest lavas has been radiocarbon-dated as about three hundred years old. Thus Taranaki might indeed be only brooding. (About fifteen years ago unexplained booming noises were heard in the vicinity of the peak, creating much uneasiness among the local residents, who feared the volcano might be awakening. However, an official investigation concluded that it was only the sound of quarry blasting some distance away, reflected back to the ground under certain weather conditions.)

The detail about Ruapehu's broken cone suggests that the legend has its basis in some real eruption of that volcano—possibly at about the same time as one of Taranaki's—which was witnessed by the Maoris before the coming of the *pakeha* (white man). Changes in the shape of volcanic summits are a common result of eruptions, the top either being blown off or collapsing into the crater, or sometimes being built up higher; the legend suggests that a collapse may have been observed in this case. Ruapehu and Ngauruhoe are still active. The area in which they are located, not far from Lake Taupo, has been set aside as the Tongariro National Park. For a long time an eruption of Ngauruhoe was taken as a command to the rather warlike tribes of the Taupo area to make war on others.

The Maoris have a tale to explain how fire came to the volcanoes and thermal areas of New Zealand [196:9]. Ngatoro was a powerful *tohunga* or medicine man, a leader of the first canoeload to arrive from "Hawaiki." One day Ngatoro climbed Tongariro, taking with him his favorite slave, Auruhoe. He warned his followers to fast until their return, to lend them strength against the high-altitude cold. When they did not return when expected, the others gave them up for dead and broke their fast. At once Ngatoro and Auruhoe on the mountaintop felt the intense cold, and both would have perished had not the *tohunga* prayed to his sisters back in Hawaiki to send fire to warm them. The sisters, powerful sorceresses in their own right, called up fire demons who immediately began swimming under water toward New Zealand. At White Island (Plate 26) they came up to get their bearings; the earth there

PLATE 26. *White Island, New Zealand, a volcano in a state of solfataric activity. In Maori legend it was a place where the demons bringing fire to New Zealand's volcanic area surfaced to look for land. (Photo by D. Callard, Tauranga, New Zealand, August 1969.)*

burst into flames which are still burning. (White Island is a volcano in the Bay of Plenty. Most of its known activity has been of the type called solfataric, emitting steam and sulfurous exhalations from several vents. Occasionally there have been mild eruptions consisting of intermittent explosive activity. The most recent of these lasted from November 1966 to April 1969 [38].) Reaching the mainland the demons continued traveling underground toward Ngatoro, and wherever they surfaced to look for him they left a thermal area. At Tongariro's summit they burst forth. The warmth revived Ngatoro, but Auruhoe had already succumbed. Ngatoro cast her body into the crater, and thenceforth the mountain was known as Ngauruhoe. (The name Tongariro is now applied to a smaller, apparently extinct volcano adjacent to Ngauruhoe.)

The eternal triangle theme also appears in connection with three much older volcanoes in New Zealand [40:101ff.]. Kakepuku and Kawa are two basaltic hills on the Waipa plain, about twenty miles south of Hamilton (see Fig. 2). The legend goes that Kakepuku fell in love with Kawa but had to defeat several rivals to win her. One of the most persistent was another volcanic hill named Karewa. Karewa put up a good fight but was finally forced to retreat into the sea, where he became what is now known as Gannet Island. Sometimes a stream of mist drifts inland from Karewa to Kawa.

Kakepuku and Kawa are eroded volcanic cones lying along a fissure on which also are located two much larger volcanic masses, Pirongia and Kariori. Gannet Island is a tiny island about twelve miles off the west coast at about the same latitude as Kakepuku and Kawa; it is not on the same fissure, but has been mapped as part of the same volcanic series. All of them have been long extinct. (In an alternate legend from the same area, Kakepuku is female, the wife of Pirongia.)

In 1886 the north and middle peaks of Mount Tarawera, in the thermal area of New Zealand, blew up violently and unexpectedly, burying three Maori villages in a rain of hot mud, stones, and ash. More than a hundred were killed. The disaster was interpreted as punishment for some transgression, and at least two legends have arisen around the event. One blames the eruption on the people of Te Ariki village, who broke a tabu of the sacred mountain by eating forbidden wild honey. Those who ate the honey were killed, while another tribe in a nearby village who did not partake of it was spared [196:142]. The other tale is more elaborate [40:145ff.]. Tamaohoi was a man-eating demon who once lived on the flanks of Mount Tarawera. Ngatoro imprisoned him in the mountain, and there he slept for centuries. One of his descendants, a *tohunga* named Tuhoto, deplored the decline in morals of his people under the influence of the white man, and finally called upon the demon to punish them. Tamaohoi burst forth from his prison and annihilated the sinners. This is an excellent modern example of euhemerism; a hundred-year-old *tohunga* named Tuhoto, who had an ancestor named Tamaohoi, was dug out alive from his buried hut four days after the Tarawera eruption, and lived for several days after [40:151].

From Mexico comes a gentler tale of volcanic lovers, in which no rival complicates the smooth course of true love as in the case of Ruapehu and Tongariro or Kawa and Kakepuku. The Aztecs believed that the two magnificent volcanic peaks which form the southeastern framework of the Valley of Mexico, Popocatepetl (the Smoking Mountain) and Ixtaccihuatl (the White Lady), were lovers who could not bear to be out of each other's sight [108:24]. The skyline of Ixaccihuatl definitely resembles the outline of a reclining woman.

The Klamath Indians of our own Northwest have a legend about volcanoes in conflict [36:53ff.]. Llao, chief of the Below World, and Skell, chief of the Above World, once battled each other from their respective positions on Mount Mazama in Oregon and Mount Shasta in California, two peaks about a hundred miles apart. They hurled rocks and flames at each other, while darkness covered the land around. The fight ended when Mount Mazama collapsed under Llao, precipitating him back into his underworld domain, where he has remained ever since, and leaving a huge hole which filled with rain to form a lake—Crater Lake. The interesting thing about this legend is that, stripped of the supernatural elements, it describes rather accurately how Crater Lake *was* formed. Mount Mazama did erupt with great violence and collapse more than 6,500 years ago, leaving the depression now occupied by the lake (Plate 27).

PLATE 27. *Mount Mazama, Oregon, at the beginning of the climactic phase of the eruption of about 6,500 years ago (above) and immediately after the collapse of its summit (below). From a series of paintings by Paul Rockwood. (From "Ancient Volcanoes of Oregon," by Howel Williams. Reproduced by permission of the Condon Lecture Series of the Oregon State System of Higher Education.)*

Crater Lake is a classic example of what volcanologists call a caldera [273].* A caldera is a large, more or less circular depression formed when a volcano collapses into the void created below it after the eruption (or subterranean withdrawal) of large quantities of material. In the Mount Mazama eruption, the surrounding countryside was blanketed by ten to twelve cubic miles of volcanic ash [272:49]. At the height of the activity, day undoubtedly must have been turned to night by the dense eruption cloud. The Klamaths were living in the area at the time, for sandals and other artifacts have been found buried in the ash. The question immediately comes to mind: If the destruction of Mount Mazama was remembered in legend, might there not have been an eruption of Mount Shasta at approximately the same time (within the same generation, say) which could have caused the two volcanoes to be associated with the same legend?

Mount Shasta has been built up primarily of lavas, but the last eruption from its summit cone produced only pyroclastic ejecta, mainly pumice, which is formed only in highly explosive activity. Subsequent eruptions (from satellite craters or fissures on the flanks) have produced mainly lava flows, the youngest of which may be as little as two hundred years old [39:29]. Thus for all we know, there could have been a fairly violent eruption of Shasta around the same time as the one which produced Crater Lake; but unfortunately no dating seems to have been done on the younger products of Shasta.

The Modoc Indians of southern Oregon and northern California seem to remember Shasta's active days. They account for the mountain thusly [36:9ff.]: The Chief of the Sky Spirits found it too cold in the Above World, so he drilled a hole in the sky with a rotating stone, and through the hole pushed snow and ice to form a mound which almost touched the sky. Then he stepped down to earth, and after creating the trees, rivers, animals, fish, and birds, he brought his family to dwell in the mountain. Sparks and smoke from their hearth fire flew out of the hole in the top of their lodge.

---

* The word comes from the Spanish *caldera* (Portuguese *caldeira*), which means cauldron; it originally was used by the natives of the Canary Islands to designate any natural depression of cauldron shape, and later was introduced into the geologic nomenclature to describe the summit depression of La Palma volcano in particular [273:240].

When he threw a big log on the fire, sparks flew up higher and the earth trembled. Eventually he put out the fire and went back to live in the sky.

Although there is no historical record of activity of Mount Hood (see Fig. 11), the geologic evidence suggests that it may have erupted as recently as a century ago. An Indian legend from the area [36:15-16] reflects a history of activity, and also an awareness that its fires may not be permanently quenched. The mountain was said to be the abode of evil spirits who, when angry, threw out fire and smoke and streams of liquid rock. In those days the Indians were all as tall as the pine trees, and their chief was the bravest, strongest, and tallest of all. One night in a dream the chief was warned that unless he conquered the evil spirits, they would flood the land with fire. So he climbed to the top of the mountain, where he found the hole leading to the home of the evil spirits. Into this hole he hurled rocks, which the spirits heated red hot and hurled back. For days the battle raged, until the chief paused to get his breath and observed that despite his desperate efforts the once green and beautiful land lay blackened and ruined. Heartbroken, he sank to the ground and was buried under the streams of hot rock. When the earth cooled and the grass grew once more, those who had taken refuge on distant mountain tops returned; but their children had starved so long that they were no longer tall and strong like their forebears. Until there comes a chief who can conquer the fire demons for all time, the people will remain stunted and weak. Sometimes, it is said, the face of the chief can be seen in shadowy profile, about halfway down the mountain on the north side.

In a Nisqually legend [36:30-31], Mount Rainier moved to the east side of Puget Sound to escape crowding when all the mountains on the Olympic Peninsula side (see Fig. 11) grew too fast. In her new location she became a monster who sucked in any creature that approached too close. Finally the Changer came, in the shape of Fox, and defied her to swallow him. As he had tied himself to a neighboring mountain with strong ropes, the monster sucked and sucked in vain, until she burst a blood vessel and expired. There have been no lava flows from Mount Rainier recently enough to have prompted the suggestion of rivers of blood flowing down the mountain, but a volcanic mud flow once poured forty-five miles down the White River valley to the lowlands west of Tacoma, and there spread out in a lobe twenty miles long and three to ten miles wide. Wood found in that mud-flow has been dated as little less than five thousand years old. Thus it is just possible that the "rivers of blood" are the memory of that event .

No land is more completely volcanic in origin than Iceland, and few countries have seen more volcanic activity in historical time. It is therefore quite surprising that the early Icelanders attached very little supernatural significance to the frequent eruptions. Of course, pre-Christian Iceland believed in the old Norse pantheon. The Eddas speak of the giant Surtur, guardian of Muspell, a flame-world somewhere to the south; one day Surtur would destroy the world with fire, allowing a newer and happier one to rise out of the sea. But in general, Icelanders have always accepted their volcanoes as a natural part of their existence and have dreaded news of an eruption purely because of the destruction it might cause to farmstead and livestock.

But not so the rest of the world! In the middle ages it was widely believed that Hekla was the principal entrance to hell. Hekla is the most feared of the Icelandic volcanoes, for by virtue of its location it menaces more inhabited districts than any other. Dormant for centuries, it erupted for the first time in the memory of man in 1104, and since then has erupted at least thirteen times. The oldest written mention of Hekla was in about 1180 by Chaplain Herbert of the Cistercian monastery of Clairvaux:

> The renowned fiery cauldron of Sicily [Etna], which men call Hell's Chimney . . . is affirmed to be like a small furnace compared to this enormous inferno. . . . Who is there so refractory and unbelieving that he will not credit the existence of an eternal fire where souls suffer, when with his own eyes he sees the fire of which I have spoken . . . ? [238:5]

Nearly four hundred years later Caspar Peucer wrote:

> Out of the bottomless abyss of Heklafell, or rather out of Hell itself, rise melancholy cries and loud wailings, so that these may be heard for many miles around. Coal-black ravens and vultures flutter about. . . . There is to be found the Gate of Hell, [and] whenever great

battles are fought or there is bloody carnage somewhere on the globe, then there may be heard in the mountain fearful howlings, weeping, and gnashing of teeth.

S. Thorarinsson, who quotes these words in his book *Hekla on Fire* [238:6], adds that traces of this superstition still lingered on into the nineteenth century, and that to this day a Swede who wants you to go to the devil may tell you to "go to Mount Hekla." In medieval Icelandic annals there are only two known references to the supernatural in connection with Hekla [238:6]. In the report on the 1341 eruption in the *Flatey Book* it is related that people saw what looked like birds, which they took to be souls flying in the fire with loud cries; and in the *Annals of the Bishops* it is said that during the 1510 eruption people thought they saw certain indications that King Hans of Denmark had landed in Hekla after his death. Of these instances Thorarinsson observes:

> I may point out that those who during the recent eruption of Hekla had the opportunity to stand for hours near erupting craters and watch fantastically shaped fragments of black lava come hurling out of the columns of smoke with an eerie hissing sound, will not think it strange that our ancestors took these for monstrous birds or the souls of the damned. . . . [Plate 28] As for the second instance, King Hans did not, in fact, die until two years after the eruption of 1510 began, and it is quite possible that the report has more in it of wishful thinking than of simple superstition. [238:6]

An old account by one of the Irish monks who roamed the North Atlantic in their skin boats describes what undoubtedly was an Icelandic eruption in the following terms:

> . . . they came within view of an island, which was very rugged and rocky, covered with slag, without trees or herbage, but full of smiths' forges . . . they heard the noise of bellows blowing like thunder. . . . Soon after one of the inhabitants came forth . . . ; he was all hairy and hideous, begrimed with fire and smoke. When he saw the servants of Christ near the island he withdrew into the forge, crying aloud: "Woe! Woe! Woe!" St. Brendan again armed himself with the sign of the Cross and said to his brethren: "Put on more sail and ply your oars more briskly that we may get away from this island." Hearing this the savage man . . . rushed down to the shore, bearing in his hand a pair of tongs with a burning mass of the slag of great size and intense heat, which he flung at once after the servants of Christ. . . . It

PLATE 28. *Explosive activity in one of the Öldugígar craters at the foot of Mount Hekla, Iceland, during the flank eruption of 1970. Activity of this type undoubtedly is what inspired the belief, widely held in the Middle Ages, that the bird-like chunks of still-soft lava were the souls of the damned hovering over the entrance to hell. (Photo by Sigurdur Thorarinsson.)*

passed them at a furlong's distance, and where it fell into the sea it fumed up like a heap of burning coals and a great smoke arose as if from a fiery furnace . . . all the dwellers of the island crowded down to the shore, bearing, each one of them, a large mass of burning slag which they flung, every one in turn, after the servants of God; and they then returned to their forges, which they blew up into mighty flames, so that the whole island seemed one globe of fire and the sea on every side boiled up and foamed like a cauldron set on a fire well supplied with fuel. All day the brethren, even when they were no longer within view of the island, heard loud wailing noise from the inhabitants thereof, and a noisome stench was perceptible at a great distance. [242:9-10].

The sights and sounds and stenches of an eruption are all captured vividly in this description, down to the hissing of red-hot volcanic bombs as they fall into the sea. But apparently an eruption was beyond the experience of St. Brendan and his crew, and thus they imagined the savage inhabitants and their forges to explain the phenomena they witnessed.

As a land of contrasts Iceland cannot be topped; the tourist brochures do not exaggerate when they describe it as the "land of fire and ice." By this they do not mean just snow-capped volcanoes —any peak high enough can have a year-round cover of snow, even on the equator. Nor is it a matter of volcanoes here, glaciers there. Active volcanoes actually exist and erupt under the ice caps. The snowy whiteness of the glacier known as Höfdabrekkujökull, an outlet of the Myrdalsjökull ice cap (see Fig. 3), hides one of Iceland's most active and most destructive volcanoes, Katla. There is a rather gruesome folktale which tells how Katla got its name [241], but to appreciate it the geologic setting must first be explained.

In the case of a subglacial volcano like Katla, there is in addition to the usual danger that good farmland will be buried under ash or lava the potentially even more disastrous possibility of a *jökulhlaup*, or glacier burst. Tremendous volumes of ice are constantly being melted by the heat from solfataras and fumaroles of subglacial volcanic fields and from occasional eruptions; this water accumulates under the ice until the pressure becomes too great and it bursts forth in a flash flood of unbelievable proportions. Glacier bursts from Katla (*Kötluhlaups* or "Katla-bursts") generally last less than a day, but they may transport enormous amounts of debris and ice and over the centuries have been moving the shoreline seaward at an accelerated rate. For instance, before 1170 Hjorleifshöfdi (see Fig. 3) was a promontary jutting into the sea, with a shallow bay to the west of it; now it is three kilometers from the shore. The name for an outwash plain, which in this and other cases glacier bursts are helping to extend seaward so rapidly, is *sandur*.

One of these Katla-bursts figures in the story of how Katla got its name: At Thykkvabaejarklaustur there was a Benedictine monastery, founded in 1186. The abbot of this monastery had a housekeeper named Katla, whom everybody feared, even the abbot himself; for not only was she very bad-tempered, but she practiced witchcraft to boot. Nobody feared her more than Bardi, the shepherd lad who tended the monastery's flock. If any sheep were missing, Katla abused the boy unmercifully. One autumn the abbot and Katla left to attend a ceremony in another district. She instructed Bardi to muster all the sheep before their return, or else! As the time drew near when they were due back and several sheep were still missing, Bardi in desperation borrowed Katla's magic breeches, whose virtue was that their wearer could run all day without tiring, and with their aid soon rounded up the rest of the flock. As Katla allowed no one to wear the breeches save herself, he tried to replace the breeches very carefully, but she quickly perceived that they had been worn and in a rage drowned the boy in a tub of *skyr* (a cultured milk product, exclusively Icelandic, which is very similar to yoghurt). Nobody could imagine what had become of the lad even though, as the winter wore on and the level of the *skyr* in the tub was lowered, Katla could be heard muttering, "Soon Bardi will appear." When it became obvious that the evidence of her crime could no longer remain hidden, Katla donned her magic breeches and betook herself off toward the glacier, and was never seen again. It was rumored that she had jumped into a crevasse, and everyone was sure of it when shortly thereafter a glacier burst roared toward the district where the monastery was located. This was immediately attributed to Katla's witchcraft. The crevasse into which she was presumed to have jumped was named "Katla's cleft," the area covered by debris from the glacier burst was known as "Katla's *sandur*," and the volcano which we now know was responsible for the flood—and for the folktale accounting for it—was appropriately given the name of the evil presence supposedly lurking beneath the ice (Plate 29).

The volcano Huzi in Japan, more familiarly known to us as Mount Fuji or Fujiyama, is considered by some to be the most beautiful peak in the world (Plate 30). Its name may have come from Fuchi (or Huchi), the Ainu goddess of fire. This sacred mountain is a perfectly symmetrical cone, at a little under 12,400 feet the highest mountain in Japan. Its last eruption was in 1707. According to a legend told me by the late Hisashi Kuno, Japan's foremost volcanologist, the mountain was built up as follows: A certain giant decided one day to fill in the Pacific Ocean. All one night he labored, scooping up bagful after bagful of earth (from some place

PLATE 29. *Looking across the outwash plain* (sandur) *toward the Myrdalsjökull ice cap, beneath which lurks one of Iceland's most feared volcanoes, Katla. (Photo by the author, August 1960.)*

PLATE 30. *Fuji-san, Japan's Eternal Mountain, from Oshiro-mura. (Photo by Hiroshi Nakano, Tokyo, February 1970.)*

such as Siberia?) and dumping it into the sea. When morning came he was disgusted to see how little progress he had made, and abandoned the whole project. He dumped the last bagful all in a heap—and that is Fuji-san.

Another version of its origin is quite different, but in this one also the mountain is said to have been formed well-nigh instantaneously [41:137]. Many years ago there dwelt on the plain of Suruga a poor woodman named Visu. One night, just as he was dropping off to sleep, he heard a terrible rumbling in the earth, and fearing an earthquake he snatched up his younger children and rushed out of his hut. What a sight met his eyes! Where there had been just an empty plain there now stood a lofty mountain from whose summit leaped flames and clouds of smoke. Bathed in sunlight the next morning, the peak gleamed like opal. Greatly impressed, Visu dubbed it Fujiyama, the Eternal Mountain. One day as Visu sat enjoying the beauty of Fuji (which was about all he did those days, much to the distress of his wife and family), he mused that it would be nice if the peak could see itself in all its magnificence. No sooner had the thought entered his mind than a great lake, shaped like the Japanese lute called a *biwa*, suddenly appeared at the foot of the mountain; and that is how Lake Biwa came to be. Actually, Lake Biwa is about a hundred and forty miles from Fuji, and another legend has it that Lake Biwa appeared at the same time that Fuji rose from the earth. It is suggested that this part of the Visu legend is an echo of some early eruption which resulted in the formation of one of the numerous small lakes at the foot of the mountain, rather than of Lake Biwa. Lakes can be created during eruptions when lava dams up streams.

Two legends concerning two different Javanese volcanoes have the same theme, the setting of an impossible task to prevent an undesirable marriage. Tangkuban Prahu (the "Overturned Proa") is an active volcano overlooking the plain of Bandung from the north. Its story runs thus [167]: Sangkuriang, a king's son, was an obstreperous little boy. One day he so exasperated his mother that she lost her temper and struck him on the head, inflicting an ugly wound. The king, who doted on his son, was so furious that he repudiated his queen and banished her to a distant part of the country. Time passed, and Sangkuriang grew up to be a handsome young man. When he came of age, his father gave him permission to travel the length and breadth of Java. In the plain of Bandung he met a beautiful woman with whom he promptly fell in love. She returned his love, and accepted his proposal of marriage. One day while caressing her lover's head she felt the wound and realized with horror that this was her own son! The marriage had to be prevented at all costs, but she could not bring herself to confess the truth. The wedding day drew nearer, but still she could think of no way out of the dilemma. Finally, on the day before the wedding, she asked that Sangkuriang prove his affection by providing a beautiful proa on which to celebrate the wedding with a great feast, together with a lake on which it could sail. Sangkuriang prayed to the helpful spirits, the *dewatas*, and they caused landslides, which dammed up the Tarum River which drains the plain; meanwhile other *dewatas* cut down a great tree and constructed a huge proa, while some busied themselves with preparations for a feast fit for the gods. The next morning the queen was dumbfounded to see that the impossible had been accomplished. In desperation she prayed to Brahma to help prevent the incestuous marriage. Brahma destroyed the dam, and the waters of the new lake gushed forth so turbulently that the proa capsized and Sangkuriang was drowned. In anguish, the queen threw herself onto the hull of the capsized vessel with such force that she broke right through and was also drowned.

And so the overturned proa stands today above the plain; the hole where the queen plunged through the hull is the Kawah Ratu, the "Crater of the Queen"; the steaming fumaroles in the crater and the frequent tremors felt on the mountain show that the heartbroken mother is still sobbing for her son. Bukit Tugul, the "Trunk Mountain" east of Tangkuban Prahu, is the stump of the tree cut to build the proa, and Mount Burangrang west of the volcano is the "Crown of Leaves" intended to be worn in the wedding festivities. This legend combines purely etiological elements, inspired by the fancied resemblance of prominent topographic features to the objects for which they are named, with what could be a folk memory of a real lake which once filled the basin. Geologic studies have shown that the Tarum River was dammed in Neolithic time by landslides from Tangkuban Prahu, forming a lake whose level originally stood at about 2,386 feet above sea level. Obsidian artifacts of the primitive inhabitants of the area have been found on the shores of that vanished lake, which must have been at least a hundred and sixty-five feet deep. As the river eroded through the loose material that blocked it, the lake level gradually sank until only a marshy plain remains today; but its former existence appears to have been remembered down the centuries in a form understandable to a people to whom tabus and ghosts and gods and spirits were much more real than down-to-earth geologic facts.

The other Javanese legend concerns Mount Bromo in East Java. As told to me by another volcanologist, Prof. Robert Decker, a powerful giant once asked for the hand of a king's daughter. The princess found the giant repulsive and her father had no desire to force her into the match, but at the same time he feared the giant's wrath. Instead of rejecting the giant flatly, therefore, the king promised him the girl on condition that he level Tengger Mountain in one night, a task which he and the princess were sure would be impossible. But the unwelcome suitor found a giant coconut shell and attacked the mountain with such vigor that it began to look as though he might succeed. But the princess was equal to the occasion; she woke up the rooster, which crowed lustily in annoyance. Tricked into thinking that dawn was imminent, the giant gave up, dropped the coconut shell, and strode away never to be seen again in that neighborhood. The partially scooped-out mountain is the Tengger caldera (Plate 31), and the overturned half of the coconut shell is Batok, a steep, gullied, extinct cinder cone next to Bromo, the active central cone in the caldera (Plate 32). This legend, obviously, is purely etiological.

North of Lake Kivu in Zaire (formerly the Democratic Republic of Congo) there is a group of volcanoes known as the Virunga volcanoes. A hero or demigod named Ryang'ombe the Ox-Eater is said to have established himself after his death in Muhavura, the easternmost of those volcanoes (Fig. 25). Frequently he fought with his enemy Nyiragongo, who originally dwelt in Mikeno. In one battle Ryang'ombe split Mikeno in two, whereupon Nyiragongo fled westward to the volcano that bears his name today. Ryang'ombe cut off the top of that peak, thrust Nyiragongo into it, and piled hot stones on top to keep him down [268:117].

PLATE 31. *Model of the Tengger caldera, Java, in the Geological Museum at Bandung. North is to the right. (Photo by R. W. Decker, 1959.)*

PLATE 32. *Looking west from the rim of the Tengger caldera. Bromo (left, steaming) and extinct Batok (right). (Photo by R. W. Decker, 1959.)*

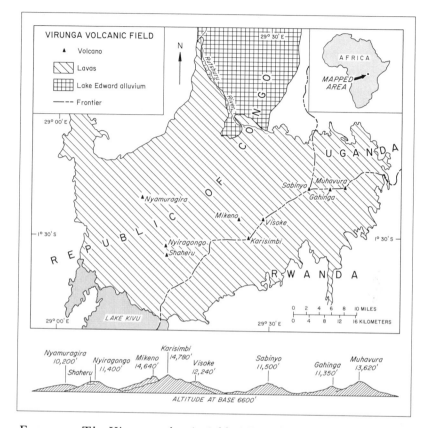

FIGURE 25. *The Virunga volcanic field of East Africa. As in Hawaii, the folklore of the tribes living in the area reflects the fact that the center of volcanic activity has migrated, in this case from east to west; only Nyamuragira and Nyirangongo are active at present. Profile at bottom from a sketch by M. Denaeyer, Brussels.*

How does this legend fit with the geologic story? The Virunga volcanoes are all young—so young that the waters of Lake Kivu once drained northward through Lake Edward to the Nile, which had to find a new outlet to the south via the Ruzizi River and Lake Tanganyika to the Congo when the chain of volcanoes was built up across their path. Nyiragongo and Nyamuragira (or Nyamlagira) are still active; the latter erupted as recently as 1957-58. The legendary battles between Ryang'ombe and Nyiragongo no doubt commemorate real eruptions. Mikeno is an old cone, however, and its double peak could be the work either of erosion or of explosion;

if the latter, it would be hard to say whether Mikeno "blew its top" at a time when the ancestors of the present tribes were there to witness it. Nyiragongo's truncated top is not the result of any event as violent as the legend suggests; it is a caldera, like that of Kilauea. Its resemblance to the Hawaiian volcano is further strengthened by the fact that there is a fire pit in the caldera floor in which a lava lake constantly boils, as Halemaumau did up to 1924. No wonder that the local tribes think in terms of an imprisoned demon!

Another volcanic peak in which a demon is believed to be imprisoned is Mount Demavend, south of the Caspian Sea in Iran (see Fig. 27, Chapter 7). Demavend, the largest single cone in the Elburz Mountains, is in an active fumarolic stage of activity, with no known eruptions in historical time. The sulfurous vapors and steam issuing from its vents have been attributed to the breathing of a demon named Biourasf; occasional rumblings are his groans. Others say that the mountain is the prison of fallen angels incarcerated by King Solomon.

The same idea of imprisoned giants or demons as the cause of various volcanic manifestations goes back at least to classical time. According to the ancient Greeks, the first children of Heaven (Uranus) and Earth (Gaea, from whom we get the "geo-" prefix in the earth sciences) were three monsters with fifty heads and a hundred hands; these represented the violent forces of nature—earthquakes, eruptions, hurricanes, thunder, and lightning. Their father hated them and imprisoned them in the earth. The next offspring were the one-eyed, man-eating Cyclops, somewhat closer to humans in proportions. After the Cyclops came the Titans, one of whom was Cronos (Saturn). When Earth appealed to the Cyclops and Titans to help release their elder brothers, monstrous though they were, Cronos responded. He ambushed and wounded Uranus, and from the blood that flowed sprang the fourth race of monsters, the Giants. Cronos ruled the universe for a long time until he was deposed by his son Zeus, in a terrible war that almost destroyed the whole universe. Zeus released the first monsters, who fought on his side with their weapons of thunder, lightning, and earthquake. Zeus learned to control the thunder and lightning and with them subdued the giants Enceladus, Briareus, and Typhon and buried them under Mount Etna. (Some versions of the myth say he buried only Typhon, others only Enceladus; and some name other volcanoes.) Their struggles to free themselves are the cause of volcanic earthquakes, and their fiery breath escaping, the cause of eruptions. The Cyclops, who had aided Zeus, were allowed to roam free over Sicily; they helped Hephaistos, the god of fire and metal, to forge Zeus's thunderbolts in a smithy located, variously, in Mount Etna or in one of several other volcanoes.

Off the east coast of Sicily at Aci Trezza, just north of Aci Castello (see Fig. 15), there is a group of rocks known as the Cyclops Rocks (Plate 33). In Homer's *Odyssey*, when Odysseus

PLATE 33. *The Cyclops Rocks at Aci Trezza, Sicily. This striking group of offshore rocks, actually the wave-eroded remnants of intrusive plugs, was thought by the ancients to be the rocks hurled by the blinded Polyphemus after the escaping Odysseus and his crew. (Photo by courtesy of* FOTO ENIT ROMA.*)*

and his crew were captured by the Cyclops Polyphemus and imprisoned in his cave, they blinded him and escaped by clinging to the bellies of his sheep. As they sailed away the enraged Cyclops hurled huge boulders in the direction of Odysseus' voice; the offshore rocks at Aci Trezza are supposed to be those very boulders, lying where they landed. The association of one-eyed giants with Etna is thought to stem from the glow in the night sky over the peak, visible far out to sea when the volcano is active. The idea of hurtling rocks likewise must have originated to explain the volcanic bombs and blocks of all dimensions that are thrown out from Etna during eruptions. The Cyclops Rocks at Aci Trezza, however, are not really volcanic ejecta, but the wave-eroded remnants of small plugs of basaltic rock which were harder than the surrounding rock; one of them still wears a cap of the marine sediments which they intruded. To the untrained eye they do look extraneous to their setting.

It has been suggested [214] that Talos, the formidable giant who patrolled the shores of Crete, may have been the personification of the volcano Santorin, the southernmost of the Cyclades islands, which "guarded" the approach to Crete from the mainland. Talos, wrought of bronze by Hephaistos, was invulnerable save for a spot on his ankle where the vein carrying his life blood was covered only by a thin membrane. He could heat himself up in the fire and kill strangers by enveloping them in a red-hot embrace. When Jason and his band of Argonauts tried to put ashore on Crete on their way home from Colchis with the Golden Fleece, Talos hurled rocks at them and would have sunk the *Argo* had not the oarsmen beat a hasty retreat. Medea cast a spell on the giant which clouded his sight, and as he was heaving up a particularly large chunk of rock to throw, he grazed his ankle on a sharp crag. As his blood flowed out "like molten lead," his strength ebbed and he fell headlong and expired. In the volcanic interpretation the hurtling rocks would be volcanic bombs, the ankle could be a subsidiary crater, the blood which flowed from the wound could be molten lava (which actually looks metallic rather than red in direct sunlight), the collapse and death of the giant could be the quieting down of the volcano after an eruption, and the red-hot embrace is all too obviously the fate of anyone approaching an erupting volcano too closely.

A Mediterranean legend from the Christian era provides an example of a geomyth which actually has contributed to the solution of a geologic problem. In his *Voyage aux îles de Lipari*, published in 1783, Déodat de Dolomieu [50:31] recorded a local tradition concerning St. Calogero, a hermit who lived on Lipari (the main island of the group called the Lipari or Eolian Islands—see Fig. 15) in the sixth century. St. Calogero is credited with having driven

away the devils who at that time were believed to be responsible
for the subterranean fires of Lipari's craters. He chased them first
to Vulcanello and then, because the inhabitants of Lipari thought
that still too close for comfort, to Vulcano; ever since, the fires of
Lipari have remained out. This legend helped the German volcanologist Jörg Keller [123] pin down the date of the last eruption
on Lipari to within fifty years, which is reasonably precise from a
geologic point of view for something that happened nearly fifteen
hundred years ago. Since a pumice tuff produced in that eruption
overlies Roman ruins dating from the fourth and fifth centuries
A.D., he knew that the eruption must have occurred after the fifth
century; but if any historical record of the eruption ever existed,
it did not survive the Dark Ages. Inasmuch as St. Calogero is definitely known to have lived in A.D. 524-562, Keller inferred from
the legend that the eruption probably occurred some time between
A.D. 500 and 550. The same legend, incidentally, faithfully reflects
the fact that volcanic activity is progressively younger from Lipari
to Vulcanello to Vulcano.

Before we leave the subject of volcano lore, we might mention
that it is to Vulcan (the Roman name for Hephaistos) that we are
indebted for the word volcano, and to the individal volcano which
bears his name for the more obscure volcanological term *vulcanian*,
which describes a particular style of activity typical of Vulcano's
eruptions. Once more, mythology has made a contribution, albeit a
slight one, to science.

7 /

The Deluge

THERE IS A SERIES OF TRADITIONS WHICH STANDS apart from all others in that these traditions appear in every part of the world—or, to be more exact, in *almost* every part of the world. These are the traditions of a great flood which destroyed either all mankind, or at least a substantial number of the earth's inhabitants. Such traditions are so widespread that many believe them to be a "racial memory" of some catastrophic inundation which affected at least a very considerable portion of the globe simultaneously. At the other extreme are those who believe that all the different flood traditions stem from a single local flood, the deluge of the Bible, the memory of which was disseminated as man migrated from the original scene. Does the geologic evidence support either of these diametrically opposed viewpoints, and if not, what does it tells us?

When geology was just beginning to emerge as a science there was no problem. The earliest geologists did not doubt that fossils in solid rocks now high above sea level, which even primitive peoples recognized as the remains of creatures which once lived in water, were left there by Noah's flood; and there still are people who believe this. It was very soon recognized, however, that rocks containing marine fossils had been laid down very slowly and over a very long stretch of time beginning hundreds of millions of years ago. The oldest recognizable marine forms date from Precambrian time, well over 600 million years ago. It also became obvious that the earth's surface has never been completely submerged at any one time. While the sediments which later were consolidated into rocks such as sandstone and shale were being deposited in one place, some other area had to be above sea level, being eroded to furnish the sediment. Only limestone can be precipitated directly from sea water (or for that matter, from fresh water), but even the most extensive limestones known, those of Cretaceous time, were not laid down everywhere. The possibility of complete submergence of the earth's surface, if it exists at all, must be pushed back into some primeval era of earth history, more than 3.5 billion years ago and perhaps earlier than life itself.

For a while early geologists still continued to think that the extensive sand and gravel deposits left behind by the continental glaciers of Pleistocene time were the result of the biblical deluge, so they dubbed those deposits "diluvium" and called the time in which they were deposited the "Diluvial." But analogies with the deposits of present-day mountain glaciers soon left no doubt that these were the work of glaciers, much thicker and much more extensive than any we know today, but different only in degree rather than in kind from the modern ice caps which cover Antarctica and Greenland. We now know that during the very small portion of geologic time which concerns mankind, the distribution of the oceans and continents and even of mountains and valleys has been pretty much the same as now, and that any relatively recent changes which have taken place have, by and large, taken place very, very slowly.

The most substantial changes in sea level during man's tenure on this planet have been those connected with Pleistocene glaciation, discussed in Chapter 3. Not only are those changes far too slow to have provoked traditions of catastrophe, but in any case they fell far short of being the kind of deep flood envisioned in most of the traditions; the highest stand of sea level during the last interglacial stage was only about a hundred feet above the present level, which still leaves a very substantial part of the world above water; also, that high stand was achieved more than a hundred thousand years ago. Not only that: as we have already seen in Chapter 3, some parts of the world have been *rising* since their burden of ice was removed, and rising faster than sea level; if the relative change of sea level with respect to land since the end of the Pleistocene period were to be remembered at all in those regions, it would be as the very opposite of inundation.

Could exceptionally heavy rains ever have produced a flood so extensive as to have drowned all the earth's low places at one time? Again the answer is a resounding no! Assuming just for the sake of argument that heavy rains fell over all the land areas of the earth at some one time, we run into serious mechanical difficulties in trying to submerge a substantial amount of land under the run-off. To begin with, the only place that water can come from is the sea, for except for an insignificant amount added from the depths of the earth through volcanoes—what geologists call "juvenile water"— the total amount of water on the earth is constant; thus if any unusually large amount of water went into the atmosphere to form rain clouds, sea level would be lowered by a corresponding amount and more land would be exposed. Then, once the moisture fell as rain, what would keep it from running back to the ocean as fast as it could? The best we can do in the way of a deluge due to simultaneous world-wide extra-heavy rains would be the serious flooding of many large rivers at the same time.

Because it is impossible to produce a truly universal flood by any normal geologic process, some highly ingenious attempts have been made to invoke extraterrestrial agencies as the cause [127, 253]. Such theories always cite folklore as evidence and bolster their arguments with incorrect or farfetched interpretations of geologic features which can be explained far more easily in terms of the normal action of geologic agencies. But the universality of flood traditions can be explained very easily without requiring a widespread flood of cosmic or any other origin, if we bear in mind that *floods, plural, are a universal geologic phenomenon.* We have seen how volcano legends of peoples widely separated in time and place have many features in common. If active volcanoes were found everywhere, volcano legends no doubt would be so common that someone would look for a universal eruption as the underlying cause. As it is, active volcanoes, and with them volcano legends, are restricted to certain belts on the face of the earth. On the other hand, there is virtually no part of the globe where there could not at some time have been a flood potentially dangerous to humans in the vicinity. A river anywhere can overflow if its waters are rapidly augmented by heavy rains, or even more suddenly augmented by the bursting of a natural dam. Even the deserts have their floods, for when the infrequent rains do fall, they commonly come as cloudbursts and there is no vegetation to retard run-off. (However, desert dwellers are too knowing to be caught in one of these very temporary but incredibly awesome "flash floods.")

There was a time not too long ago (geologically speaking) when the world's climate was generally wetter than now. When glaciers covered northern North America and Eurasia, precipitation was heavier in the areas outside as well as within the regions of snowfall; rivers were larger, and many huge lakes existed beyond the ice front. Great Salt Lake in Utah is the remnant of Lake Bonneville, an ancient body of fresh water which once filled part of the Great Basin; several lakes in the Nevada desert, including Pyramid Lake and Walker Lake and recently dried-up Winnemucca Lake, are remnants of ancient Lake Lahontan. Wave-eroded shorelines, raised beaches, delta deposits, and other topographic features associated with different levels of those lakes are conspicuous—at least to the trained eye—as one travels across Utah and Nevada. As the glaciers melted, even bigger lakes were formed at the ice front. The Great Lakes of North America are now only a fraction of their former size, and there was an enormous Lake Agassiz, of which Lake Winnipeg is the largest remnant. The number of smaller lakes that were impounded temporarily by tongues of ice is impossible to estimate, but there must have been thousands of them at different times and in different locations.

When the ice dams impounding such lakes failed, they often must have failed suddenly, and there must have been many local floods which could have wiped out Indian villages farther downstream, in much the same way that an Italian village below the Vajont Dam was wiped out in 1963* but on a much larger scale. The results of meltwater floods of stupendous proportions are illustrated by the "channeled scabland" topography in the state of Washington, where extensive areas of the Columbia River basalt plateau were swept bare. The outstanding feature of that region is the series of interlocking coulees (valleys which are dry most of the time) cutting the plateau into a maze of buttes, mesas, and larger tracts. Dry Falls in Grand Coulee (Plate 34) are four hundred feet high. The Columbia River scablands were presumably scoured by

* In that case the water behind the dam sloshed over it in one huge wave as the result of a landslide into the lake; the consequences to those living in the village downstream, however, were the same as if the dam itself had given way.

PLATE 34. *Dry Falls in Grand Coulee, formed in the tremendous deluges which created the "channeled scabland" topography in the plateau basalts of eastern Washington. (Photo by the author, August 1959.)*

one or more floods caused when very large periglacial lakes were suddenly drained, and one probable source of the water was ancient Lake Missoula, created when Clark Fork was dammed by an ice tongue (Fig. 26) [245:454]. If by any chance the human memory goes back several thousands of years—as the Klamath legend of the creation of Crater Lake reported in the last chapter certainly seems to suggest—then conditions existed until well after ten thousand years ago, in North America at least, when floods resulting from the sudden draining of lakes dammed by tongues of the retreating glaciers would have been common and often serious, even if not quite on the scale of those which produced the channeled scablands.

At all times up to and including the present, there has been one source of frightful flooding of coastal areas in all parts of the world—but particularly on all Pacific shores—which would be particularly memorable to those fortunate enough to survive: tsunamis, or seismic sea waves. Although by no means universal indi-

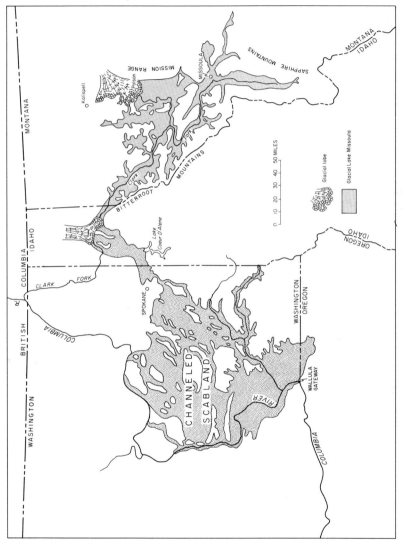

FIGURE 26. *"Lake Missoula," which drained suddenly when the ice tongue damming Clark Fork gave way and created part of the "channeled scabland" topography. Post-glacial deluges of this kind could have inspired some North American Indian flood legends. (From Regional Geomorphology of the United States, by W. D. Thornbury [245]. Reproduced with the permission of John Wiley and Sons.)*

vidually, tsunamis can and often do wreak havoc in very widely separated places, within hours. Since tsunamis are important not only as possible sources of flood legends but also in connection with matters of paramount importance in the chapters yet to come, a detailed look at these formidable waves is necessary at this point.

Tsunamis as a rule are associated with submarine earthquakes. They are generated either directly, if earthquake faulting displaces the sea floor, or indirectly, by submarine landslides, mudflows, or slumping triggered by an earthquake [176]. Occasionally they can result from a submarine eruption, if an underwater explosion displaces substantial amounts of water. Sea-floor displacements involved when a caldera collapses on the sea floor likewise can produce tsunamis, and if the caldera is a large one and its collapse sudden, as in the case of the Krakatau eruption of 1883, the resulting tsunami may be of stupendous proportions. Also in Krakatau-like eruptions, huge waves may be generated when incalculable amounts of volcanic tephra suddenly fall on the sea surface after a major explosion [254:409].

The propagation of a tsunami is an extremely complicated matter. The speed with which one travels depends on the depth of the water; speeds up to three hundred to five hundred miles per hour in the open sea are common. Because the speed depends on depth (the greater the depth, the faster the wave) and because the ocean floor is anything but smooth and flat, the wave front soon becomes highly irregular; also, the height of the wave diminishes rapidly as the wave front spreads out, roughly in proportion to the square root of the distance traveled. Thus before a tsunami has traveled very far from its source, it usually becomes not one tremendously high wave, but a serious of tremendously *long* waves— as much as one hundred to four hundred *miles* from crest to crest, but not more than a very few feet high! Ships at sea do not even notice their passing, so gradual is the rise and fall. But when such volumes of water approach a shelving shore, they pile up and can do frightful damage even thousands of miles from their source. Scientists in many countries are working to perfect tsunami warning systems, for as mentioned in Chapter 5, tsunamis are potentially the most serious consequence of earthquakes. Most of the 2,000 deaths in the 1960 earthquake in Chile were due to the tsunami that was generated [53]; and in addition, 61 people were killed by that same tsunami in Hilo, Hawaii, and 180 in Japan. Fortunately only a very small percentage of earthquakes cause the sea-floor displacements necessary to generate tsunamis, and caldera collapses on the sea floor are exceedingly rare.

Usually, but not always, the first visible sign that a tsunami is approaching is a withdrawal of the sea to far below the low tide mark. Sometimes the water returns within a few minutes, and sometimes it may retreat a few miles out to sea and return as much as a half hour later. Most people picture a tsunami as a gigantic breaker racing across the sea and looming horribly over the shore before it breaks. Actually, more often than not the water comes in either as a solid wall or, quite commonly, like a fast-rising tide. Anything floatable, including sizable ships, may be picked up and swept far inland. Solid objects weighing tons may be tossed about like pieces of wood; in the tsunami from the 1960 Chilean earthquake, large fishing boats at Ofunato, Japan, were lifted over a pier eight feet above water level and deposited a hundred and fifty feet inland amid the wreckage of houses [111:44], and in 1946 a tsunami left a block of coral weighing several tons on a pier at Mahukona harbor, Hawaii. Even if the inrush of water does not cause great devastation at a particular place, tremendous damage can be inflicted when the water drains back to sea, undermining foundations, uprooting trees, and carrying anything loose, including people, out to sea. In the village of Tjaringin on the Java coast (see Fig. 32, Chapter 8), waves caused by the Krakatau caldera collapse pardoxically were responsible for fires, for when houses were torn from their foundations by the water, overturned lamps set their contents alight [254:46].

At any given place the configuration of the shoreline and of the sea bottom, together with the local topography, will have as much or more to do with the height to which a tsunami will rise on that shore as does the initial height of the wave at its source [119]. Reefs or offshore islands may serve as protection, whereas in a funnel-shaped bay or at a river mouth the water may rise to fantastic heights. Each tsunami is unique. It has been noted in Hawaii that a particular tsunami might be more severe in one locality than another, while the effects of the next one, presumably approaching from a slightly different direction, might be just the reverse at the same two places [62]. Which wave of a tsunami train will be the highest also varies. In the one which reached Hawaii after an earthquake in the Aleutian Islands in March 1957, the third crest was the highest (nine to ten feet of run-up); in the tsunami from the Chilean earthquake of 1960, the first wave to reach Hilo was four feet above mean sea level, the second nine feet, and the third thirty-five feet; but elsewhere on Hawaii maximum heights ranged from two to seventeen feet [62]. The tsunami from the great Alaskan earthquake of March 1964 did damage along the California coast, particularly at Crescent City, where several lives were lost because of popular ignorance of the habits of tsunamis; many people who had evacuated the danger area began to return after the first and second crest had passed, but it was the third and fourth, both twelve feet high, which swept into the city. At San Francisco, an estimated ten thousand people jammed the beaches at the critical time; had a major wave, similar to the third and fourth at Crescent City, struck that particular stretch of coast, all would have perished [136].

All in all, then, from the purely geologic point of view we should *expect* independent flood traditions to have arisen almost anywhere in the world at almost any time, engendered by flood catastrophes stemming from perfectly natural causes, and of all the possible causes of floods, only tsunamis are capable of giving rise to flood legends in widely separated places at the same time. Although many different floods are required to account for the many traditions known, there is no reason to be surprised that flood traditions from all over the world may bear notable resemblances to one another. For when we come right down to it, there are only two basic ways in which people can survive a flood: by getting above it, or by riding it out on some floating object. Thus there are legends in which the survivors take to high ground or climb exaggeratedly tall trees, and there are legends in which the survivors float to safety in an ark, a canoe, a chest, or what have you. In most flood traditions a vessel is the means of salvation, and that too is not surprising, in view of the fact that the water depth is often exaggerated to the point where everything is submerged and there would be no other way to account for anyone's being saved to carry on the human race. Exaggeration likewise tends to reduce the number of survivors toward the apparently irreducible minimum of the one man and one woman needed to repopulate the world (but some legends manage to get by with even less). And finally, need we be surprised if some independently generated legends lay the blame for the disaster on somebody's misbehavior? Remember, in the preceding chapter, how the Maoris attributed the Tarawera eruption to the breaking of tabu by the victims? Nevertheless it is undeniable that many flood traditions in widely separated parts of the world do show similarities in detail, highly reminiscent of the biblical deluge, which cannot be explained entirely by the general similarity of floods and the general similarity of human reactions to floods.

There are only two ways in which the story of Noah's flood, whatever its local source, could have been spread around the world: by diffusion, as the people to whose culture it originally belonged migrated to new lands, or by transmission, which requires contact between at least one narrator and one listener from different cultures. Flood traditions are found throughout the western hemisphere from Alaska to Tierra del Fuego. In the extreme diffusionist point of view this constitutes evidence that the Indians of North and South America are descendants of one of the lost tribes of Israel, who brought the story of Noah with them as they migrated across Asia and into North America via Bering Strait and on down through South America. But while anthropologists do believe that man reached the Americas by way of Bering Strait, the waves of migration took place long before Noah's prototype existed. So that brings us to transmission and its corollary, syncretism (the fusion of elements from independent traditions). If all the biblical parallels in New World flood traditions are the result of cultural contact, then either that contact was somehow established long before the first missionaries are known to have reached the western hemisphere, which is unlikely, or else all such parallels must date from after the time of the first missionaries.

A highly illuminating example of how a legend can be transferred from one culture to another literally overnight was related by Alice Lee Marriott in a *New Yorker* article some years ago [52]. When she was collecting the folklore of a South Dakota tribe, she was challenged one day by the old Indian who was her informant to tell him one of the tales of her people. She thereupon related the story of "the Brave Warrior and the Water Monsters" —Beowulf. Few changes were necessary; it was "all within the patterns of legendary behavior, which the old man could understand, and I reflected that there might be more to this universal-distribution-of-folklore than I had realized." A little later she heard him relate the story to an audience of his people, "and I must admit the old man made a better story of it than I did. A born, creative story teller, he added bits here and there to round the tale out and make it richer. So must the story of Beowulf have gone, many centuries ago, from hearer to hearer, improved and embellished until at last it was written down." The punch line of her article told how a few years later in an ethnological journal she came across a paper entitled "Occurrence of a Beowulf-like myth among North American Indians," published by a graduate student who, in violation of an unwritten law among ethnologists, had been using the same informant.

With this illustration in mind, it seems quite natural that certain details of the biblical flood story should turn up all over the world. For more than nineteen centuries missionaries have been carrying it to every corner of the earth. The story of Noah is one of the most colorful of all the Bible stories, and it is also one whose moral is particularly obvious and therefore most likely to be emphasized. Moreover, it should have made the most impression precisely among those peoples who already had a flood tradition with which it could be fused. Missionaries have always been among the first to brave the wilderness to bring the Gospel to primitive people, and in many instances they were the first to take down the legends of the people among whom they worked. In other cases, however, the legends were collected by ethnologists and others who came well after the missionaries. Because it often was the missionaries who first devised written forms of obscure languages, it is imposible to prove whether a flood story really predates the missionary influence or whether it is just Noah being given back with local color, like Beowulf in South Dakota. Only one very equivocal instance of pre-missionary documentation is known (which will be discussed subsequently), but at least one instance has been confirmed where Noah was given back in the same way as Beowulf: A missionary named Moffat related, in a book published in 1842 [2:51-52], how he had never found a flood legend among the South Africans until one day a Namaqua Hottentot told him one. Suspecting possible missionary influence, he questioned the man closely but was assured that it was a tale of his forefathers, and that the Hottentot had never even met a missionary before. But later, when Moffat was comparing notes with another missionary, he learned that the other had indeed told the story of Noah to the very same Hottentot.

The British anthropologist Sir James Frazer [63], and others before him, notably the German geographer and anthropologist Richard Andree [2], compiled flood legends from all parts of the world and examined them for evidence of local origin versus transmission. To try to cite them all would fill a sizable volume and be very repetitious withal. So let us examine just a few typical examples from all over, in the light of their geologic setting. After that the reader can better form his own conclusions as to the origin of the ubiquitous flood traditions.

The oldest known flood story is that of Noah, the origin of which can be traced back to Sumeria. The Bible story is too well known to repeat here. What is not so well known is that the version in Genesis was compiled by some unknown editor from two distinct and not entirely consistent narratives [63:125ff.]. One of these comes from the Yahwist (Jehovist) document (J) and the other from the younger "priestly" source (P). The sending out of birds to see if the waters had subsided and the offering of a sacrifice by Noah are peculiar to J; the detailed instructions for building the ark, Mount Ararat as the resting place, and the rainbow of promise are peculiar to P. In J the flood culminates in forty days, in P, in a hundred and fifty. In J the animals take a week to embark, in P apparently only a day. And in P the flood is caused by an uprush of subterranean waters in addition to rain.

The Babylonian version is virtually identical except that the name of the chief character is Utnapishtim. The story of Utnapishtim [63:107ff.] is incorporated in the Gilgamesh epic, recorded in tablets unearthed at Nineveh in the library of Ashurbanipal (668-633 B.C.). Enough bits and pieces of older versions have been found in different places to prove that Ashurbanipal's version in turn is based on a Sumerian story which goes back to about 3400 B.C., in which the hero is called Ziudsuddu or Xisuthrus. Utnapishtim was a good man who was warned by the sea god Ea that the world was destined to be destroyed by a flood in punishment for the wickedness of mankind. As instructed, he built a boat into which he repaired with his family, skilled artisans, and animals. After seven days of tempest their vessel grounded on "Mount Nisir." Utnapishtim sent out a dove, which found no land and returned; then a swallow, which also returned; and finally a raven, which did not come back. Upon disembarking Utnapishtim sacrificed to the gods, who "smelt the sweet savor," promised that there would be no repetition of the deluge, and ultimately took Utnapishtim to dwell among them.

At first it was thought that the Hebrews might have learned this story when they were taken in captivity to Babylon by Nebuchadnezzar (605-562 B.C.), but the Genesis story in its oldest form, in the Jehovist document, is now believed to have been written in the eighth or ninth century B.C. Another suggestion is that they could have heard it from the Canaanites. But it seems most likely that the Patriarchs brought the story along when they migrated from Mesopotamia [18:40].

Attempts have been made to discredit Mesopotamia (Fig. 27) as the source of the biblical flood tradition, on the grounds that the rainfall is not heavy enough there to cause floods. However, the rains that cause a river to flood in its lower course can fall anywhere in its catchment area, and the Tigris and Euphrates are very long rivers indeed; there is also archeological evidence that floods *have* occurred there not once, but many times. A ten-foot layer of flood silt has been found at Ur in the Obeid level, indicating a flood during the fourth millennium B.C. [18, 276]; at Kish there is evidence of a flood which occurred considerably after 3000 B.C.; at Fara there is a two-foot layer of alluvium representing a flood which came some time after the one at Ur but before the one at Kish; and at Nineveh there is a layer six to seven feet thick which could be of the same age as the one at Ur, or nearly so [18:34-35]. These layers constitute a record of perfectly expectable, more or less local floods of the Tigris or Euphrates or both at once.

The Viennese geologist Eduard Suess [229:72] proposed in 1904 that flooding in Mesopotamia might have been compounded to more catastrophic proportions if a typhoon drove the shallow waters of the Persian Gulf inland upon the delta, backing up the

FIGURE 27. *Sketch map of Mesopotamia, birthplace of the Hebrew-Babylonian flood tradition.*

flow of rivers already swollen. The recent typhoon disaster in East Pakistan emphasizes the inherent plausibility of such a suggestion. Suess further believed that an earthquake could have been responsible for the idea of the bursting out of subterranean waters as an additional cause of the flood. The area is seismically unstable, and in the alluvial plains of great rivers water frequently spurts out in great fountains when the ground is compressed by earthquake stresses (such jets of subterranean water were observed in the New Madrid earthquakes of 1811-12 [65:76], among others). Frazer [63:358] points out that this detail, lacking in the earlier Jehovist and Sumerian versions, appears to be a later embellishment of the flood story. It seems to me that this strengthens the possibility that it originated from some observation of waters spouting from the ground during an earthquake, for it then removes the necessity for the overly coincidental occurrence of an earthquake simultaneous with the original flood.

The thickness of flood silt in the excavations in Mesopotamia does not prove that the waters stood deep and for a long time. It is not the depth of the water but its velocity that is the deciding factor in sedimentation. Fast-flowing water does not deposit sediment; on the contrary, it scours and erodes, picking up anything loose and carrying or rolling it along. The moment the current is checked for any reason, the particles fall to the bottom; first the heaviest, and as the current slackens, progressively finer particles fall. Local obstructions, such as buildings or walls, may either speed up the current or impede it. If the flow is constricted by being forced around or between objects, its velocity is increased and so is its capacity to scour, and the obstruction may be undermined or swept away; but if the obstructions are so placed as to create a backwater, a sizable load may be dumped in the quiet spot while other places are receiving no sediment at all. This can explain why the alluvial layer laid down in the flood in the fourth millennium B.C. is not found in all the pits that were dug through the early strata of occupation at Ur [18:35].

This flood of the fourth millennium may or may not have been the specific flood commemorated in the Utnapishtim-Noah tradition, though many, including myself, believe it was. Whichever flood was responsible, it must have been confined to the lower Tigris-Euhprates basin; but particularly if both rivers rose at once, a very considerable expanse of the low flat delta area would have been inundated, enough to have constituted the whole world of the inhabitants of Ur and other cities on the plain.

Best known to most of us after the Babylonian-Hebrew flood tradition is that of classical mythology, Deukalion's deluge. Of the several Greek flood traditions, it is the only one in which the flood is said to have been worldwide. Deukalion, son of Prometheus, was a king of Thessaly. When mankind fell into evil ways, Zeus decided to destroy the world. Prometheus warned Deukalion, who was a good and pious man, and advised him to build a large wooden chest and stock it with provisions. Nine days and nights it rained, and the waters rose so high that only the top of Mount Parnassus (see Fig. 28) stood above the flood. Deukalion and his wife, Pyrrha,

FIGURE 28. *Greece and the Aegean Sea, showing the location of places mentioned in various classical myths. The eruption of Santorin in the fifteenth century* B.C. *may have been the origin of several apparently unrelated myths and traditions, and has been proposed as the cause of the sudden demise of Minoan civilization.*

floated safely in their chest, which came to rest on Parnassus as the waters subsided. As soon as they disembarked they gave thanks for their deliverance and prayed to Zeus for help in their loneliness. Zeus commanded them to cast behind them the "bones of their mother." Interpreting this to mean rocks, the bones of mother Earth, Deukalion and Pyrrha cast behind them stones, which turned into men and women. Deukalion and Pyrrha had a son whom they named Hellen, and through him they became the ancestors of the Greeks (Hellenes).

Deukalion's flood was accepted as historical fact by the Greeks, including Aristotle. There apparently was at least one king by that name. A marble pillar found on the island of Paros gives a list of the kings of Greece and the dates of their reigns, and according to this chronicle, Deukalion's deluge occurred in about 1539 B.C. [63: 149]. However, the Parian marble's dates for early events are somewhat higher than those reckoned from extant genealogies, according to which Deukalion lived about two generations later and his flood occurred in about 1430 B.C. [166:326]. The Egyptian historian Manetho stated that Deukalion's deluge occurred in the reign of Tuthmosis III (1490-1439 B.C.). About the middle of the fifteenth century B.C., or possibly earlier, there was a Krakatau-like eruption of the volcano Santorin in the Aegean Sea (about which we will have much more to say in the next chatper). At the end of that eruption the volcano collapsed to form a caldera, and that collapse could have generated one or more tsunamis, possibly far bigger than any ever generated in the Mediterranean area in the more normal way by earthquakes. The possible dates for Deukalion and for the eruption are sufficiently close, in our present state of knowledge, that the proposal (first offered by A. G. Galanopoulos [68, 73]) that the legend or myth of Deukalion's deluge was a consequence of that catastrophe appears very plausible. In this light it appears particularly significant that Andree [2:40] states that in an early version of the myth the flood is said to have come from the sea ("*Meerflut*")—and what else could that mean but a tsunami?

Later versions of the Deukalion story include details that closely parallel the Hebrew-Babylonian flood story. In the course of time the sea flood became nine days and nights of rain, the chest became an ark, animals were included in the passenger list, and Deukalion sent out a dove on successive occasions to see if the waters had receded [63:153-156]. Thus the traditions of two different places, based on floods centuries apart, merged into what is essentially the same story. One of the differences between the Greek and the Hebrew flood traditions is that Deukalion and Pyrrha were furnished with an unspecified but presumably not small number of companions, sprung from "the bones of the earth," to help them re-populate the world; but naturally, if the Greeks believed that their flood had happened less than a thousand years before, they needed more than one family of survivors to regenerate a population equal to that of the world known to them, in the time elapsed since the disaster.

There is considerable lack of agreement concerning Deukalion and the characters associated with other Greek flood traditions. Frazer, who of course was unaware that there might have been a real geologic event at about the right time which could have caused a serious inundation, and who apparently was not impressed with Andree's probably closer-to-the-mark suggestion that Deukalion's deluge might have been the tradition of some earthquake-generated tsunami (although Frazer gives full credit to tsunamis as a possible source of some flood traditions [63:347]), called the Deukalion legend a myth of observation, made up to account for the spec-tacular Vale of Tempe (see Fig. 28) [63:174]. The ancient Greeks assumed that a vast lake was once contained within the circle of the Thessalian mountains, and that the gorge was created when the waters burst out suddenly. (In fact, the gorge is the result of normal erosion processes and the lake a product of the imagination.)

J. V. Luce [143:145] and others prefer to link Deukalion's deluge to flooding of the Copais Lake basin (see Fig. 28), a poorly drained marshy lowland (now drained and under cultivation), which was subject to flooding whenever the Cephisus River was swollen—according to Luce, possibly swollen by excessive rains caused by the Santorin eruption (which is plausible), and the out-lets of the basin blocked as a result of "earthquake damage asso-ciated with the Thera eruption" (which, for reasons discussed in the next chapter, is not plausible). Both Frazer [63:161] and An-dree [2:40] linked the Copais Lake flooding with Ogyges, a king who is said to have founded the city of Thebes in Boeotia. Ogyges' flood, the best known Greek flood tradition after Deukalion's, was not as widespread or as serious as Deukalion's, and the Greeks—to whom it too was an historical fact—believed it to have been earlier. Of course there is the possibility that Deukalion's deluge was in-spired both by coastal inundation by a tsunami *and* by flooding of a poorly drained area like the Copais Lake basin; in this case we can have it both ways, and surely the combination of the two would be more likely to have produced a tradition of worldwide flooding than either alone.

The third greatest flood in Greek tradition is the one associated with a king Dardanos of Arcadia, who was driven to Samothrace by floods in his homeland. Frazer attributed the Dardanos legend to some real local flood in the vicinity of the king's birthplace, Pheneus, in an area which, like the Copais Lake area, frequently suffered from flooding [63:163]. There are many other Greek flood traditions concerning purely local floods, some of which point rather clearly to a tsunami as their starting point. A number of these could also have been inspired by a Santorin tsunami in particular; but those can be left to a later chapter, as here we are concerned mainly with traditions of fairly widespread inundations.

Outside of Greece, flood legends are surprisingly rare in Europe [63:174]. There is one from Wales, one from Lithuania, two in Norse mythology; a gypsy legend from Transylvania involving a fish, which may have been derived from the Indian legend (q.v.) [63:177]; and one from the Voguls, a tribe living on both sides of the Ural Mountains, which attributes the flood to rain after a prolonged drought [63:178-179]. The Welsh story [63:175] tells how all of Britain was inundated when the lake of Llion burst its bounds; in its details, it is rather obviously the Bible story transplanted to a local setting. The Lithuanian legend [63:176] is more elaborate, and has elements in common both with the Bible story and with Deukalion: When the supreme God looked down from heaven and saw war and injustice among men, he sent two giants, wind and water, who laid waste to the sinful earth for twenty days and nights. When next he looked down he happened to be eating nuts, and he dropped a nutshell, which landed on the very mountain top where the animals and a few human couples had taken refuge. They all climbed into the nutshell and floated safely on the waters. After God caused the storm to abate and the waters to recede he divided up the survivors, one couple to a region. The pair assigned to Lithuania were already old, and they were lonely. So God sent them the rainbow as consolation, and bade them jump over the bones of the earth. Nine times they jumped, and each time another couple sprang up, who become the progenitors of the nine Lithuanian tribes.

No biblical influence is apparent in the Norse mythology. One of the myths is set in the days before mankind, in the time of the giants. When the evil Ymir was slain by Odin and his brothers Vili and Ve, the blood gushed forth in a flood which drowned all the frost giants except Bergelmir and his wife, who escaped to Jötunheim and founded a new race of giants [63:174]. The other Scandinavian myth involving a flood is that of the *Ragnarok*, the "Twilight of the Gods" immortalized in Wagnerian opera. It is not clear whether this debacle is supposed to have happened in the past, or is still in store like the Judgment Day of Christianity. As described by Bulfinch [24:245-246] it was yet to come that all visible creation would be destroyed. The earth would tremble, the sea leave its basin, the heavens tear asunder, and men perish in great numbers. Then the wolf Fenris would burst the chains which bound him, the serpent Midgard would rise from the sea, and their father Loki would break free of his bonds and join the enemies of the gods. Led by Surtur, the flame giants of Muspelheim would rush forth. In the ensuing battle the gods and their enemies would all be slain except for Surtur, whose flame would consume the universe. Afterward a new heaven and earth would rise from the sea. Other versions use the past tense [91:329], and mention that a few of mankind survived the holocaust by hiding within the great ash-tree Yggdrasil, which upholds the universe and which was not consumed.

The flood traditions of Asia are very diversified. The Persian myth [2:13-15] from the Bundahish, one of the later scriptures of the Parsees, tells how in earliest times the earth was full of malign creatures fashioned by the evil principle Ahriman. The angel Tistar (the star Sirius) descended in three different forms successively—man, horse, and bull—and in each form he caused it to rain for ten nights and ten days. The first flood drowned all the evil creatures, but the seeds of evil remained and poisoned the earth. Before Tistar could return (in the form of a white horse) to send a second cleansing rain, he had to do battle with the demon Apaosha, who appeared as a black horse. With the help of Ormuzd, who blasted the demon with lightning, Tistar finally prevailed. Upon being struck, Apaosha gave a fearful cry which can still be heard in thunderstorms. The poison washed from the earth by the second rain made the seas salty. The waters were driven to the ends of the earth by a great wind, where they created three great seas and twenty-one lesser ones. This myth has no recognizable resemblance to the Hebrew tradition and appears to be essentially an etiological myth explaining a number of natural phenomena.

The Hindu flood tradition is quite different [2:16ff.]. In its earliest version it tells how a man named Manu found a little fish in the water brought for his morning ablutions. The fish said, "Protect me and I will save you. A great flood is coming which will destroy all creatures." "How can I protect you?" asked Manu. "Put me in a jar until I am bigger. When I outgrow the jar, dig me a pond. When I am too big for the pond you can put me in the sea, as I will then be too large to be eaten by other fish." This Manu did. On the advice of the fish he built a vessel, and when the flood came, embarked in it. Of all the creatures on earth, only he was saved. The human race was regenerated with the help of a woman whom the gods fashioned out of the sacrifice Manu offered after his deliverance.

The legend of Manu is post-Vedic, there being only obscure references in the Vedas that might relate to it. It first appears in the Satapatha Brahmana, which goes back to about 600 B.C. However, some think it might be older than the Babylonian legend, although there is no reason to believe that it was necessarily the source of the latter [230:viii]. It has been suggested that in view of the geographic setting, the bursting of a landslide-dammed lake in the Himalayas could be a possible factual basis for the flood tradition in Sanskrit literature [230:iv].

In later versions Manu is not merely a common man, but a great seer or a king. As in the case of the Deukalion myth, possibly Semitic elements appear after a while. In the Matsya Purana, dating from 320 A.D., Manu takes all living creatures and the seeds of plants into his "ark" [63:189].

Kashmir has an obviously etiological myth similar to that attributed to the Vale of Tempe in Greece. The famous Vale of Kashmir, surrounded by high mountains, is said to have been occupied in fomer times by a lake, but Vishnu created an opening in the mountains near Baramulla through which the lake was drained, leaving the Happy Valley which was then populated by the descendants of Ksayapa, an uncle of Brahma [2:25].

The Buddhist traditions of China and Japan contain no mention of a universal flood. There is a Chinese legend which concerns the flooding of one particular river. Frazer [63:214] attributed it to an historical attempt to control the waters of the Hwang Ho (Yellow River), called "China's Sorrow" because of its all too frequent flood disasters. In the reign of a mythical sovereign named Yao, there was a terrible flood lasting twenty-two years. Yao called upon one Kun to deal with the situation. Kun struggled unsuccessfully for more than nine years, trying to contain the waters by means of dikes. His son Yu carried on the work using a different approach, attempting to divert the river into new courses, and after thirteen years he succeeded. Many bits of geographic folklore recall the labors of Yu; the famous Lung Men Gorge (Gate of Dragons), for instance, is supposed to have been cut by him to divert the Hwang Ho to the sea.

A completely different and rather imaginative Chinese flood story relates that a queen or goddess named Nu Kua fought with the chief of a neighboring tribe. Chagrined at being defeated by a woman, he beat his head against the Heavenly Bamboo and knocked it down, tearing a hole in the Sky Canopy. Floods of water poured out, but Nu Kua gathered stones of five different colors, ground them to powder, and mixed a plaster with which she patched up the rent and stopped the flood [269:225].

Flood traditions are lacking in semi-arid Central Asia, which is hardly surprising. Southeast Asia does have flood traditions, many of them showing biblical influences. One of the more interesting is that of the Lolos, an independent, literate, aboriginal race living in southwestern China. The Lolos believe in patriarchs living in the sky who once dwelt on earth and attained incredible ages, much greater even than Methuselah. When mankind became wicked one of these semi-divine patriarchs, Tse-gu-dzih, sent a messenger to earth requesting some flesh and blood from a mortal. Only one person could be found who would give any, a man named Du-mu; so Tse-gu-dzih locked the rain gates and let the waters pile up to the sky. Du-mu was saved in a hollow log, together with his four sons and several otters, wild ducks, and lampreys. From the four sons (who their wives were is not specified) are descended the literate peoples of the world, while the rest of mankind are the issue of wooden figures fashioned by Du-mu to repopulate the earth. The Lolos observe a sabbath (every sixth day, however) on which they do no work. This custom, together with the traditions of the patriarchs and the flood, has been ascribed to the teachings of Nestorian missionaries, the first of whom is said to have reached China in A.D. 635 [63:212-214].

To the Benua-Jakun, an aboriginal tribe in the Malay state of Johore, the ground is merely a skin covering a water-filled abyss [63:211]. Once upon a time the god Pirman broke this skin and all the world was drowned except for a man and a woman whom he had sealed up in a wooden vessel. When the vessel came to rest after the flood, they gnawed their way out. The woman conceived in the calves of her legs, and from the right leg a male child was born and from the left, a female. All mankind is descended from that boy and girl.

Very conspicuous by its absence is an Egyptian flood legend; but likewise conspicuous by their absence in Egypt are disastrous floods. Every year the Nile overflowed its banks gently and predictably, leaving behind a life-giving deposit of fine silt to replenish the soil. Lean years might have ensued when the waters fell short of the average, and extra-high waters might conceivably have caused some inconvenience, but the annual flood could never have been anything but benign on the whole. Its *failure* to materialize would have been the disaster to commemorate in legend. The other main rivers of Africa also have an annual rise which, being predictable, is not calamitous.

The only legend from southern Africa involving any sort of inundation is not a typical deluge tradition at all, but one which seeks to explain the origin of a particular lake, Lake Dilolo on the boundary between Zaire and Angola; it properly belongs in Chapter 4, Landform Lore, like some others already cited in this chapter. According to this tale [2:49-50], a chieftainess named Moena Monenga once sought food and shelter in a certain village. Not only was she refused, but when she reproached the villagers for their selfishness they mocked her and said, in effect, "What can you do about it?" So she showed them; she began a slow incantation, and on the last long-drawn-out note the whole village sank into the ground, and water flowed in to fill the depression. When the chieftain of the village returned from the hunt and saw what had happened to his family, he drowned himself in the lake in despair. This tale was collected by Livingstone, and was the only one he encountered in all his years of missionary work which had any resemblance to a flood tradition. He believed that the name of the lake came from the word *ilolo*, which in the local language meant "despair."

The lack of flood traditions in Egypt and the rest of Africa has been a definite stumbling block to theories requiring tremendous volumes of water sloshing over the whole face of the globe as a result of cosmic collisions. One little-known work which postulates impact with a giant meteorite about 11,500 years ago states, rather feebly, that "Egypt happened to be fortunately located with regard to the geological and tidal effects of the collision which produced the deluge" [127:249]. I find it hard to understand how Egypt could have been "fortunately located," while just across the Mediterranean the meteor-impact-generated flood was supposed to be causing the deluge of Deukalion, which is cited as evidence in favor of the same catastrophe! The interior of Africa is said to have been "even more fortunately located than Egypt, with higher elevations and closer to the point of no distortion of the geosphere" [127:249]; but then in the very next sentence the authors cite a different version of the Lake Dilolo legend which tells that a great wave *did* cross the country, leaving the lake in a depression. Again it is hard to follow the reasoning which on the one hand attributes the lack of African flood traditions to the lack of the floodwave there, yet offers folklore "evidence" that a gigantic wave washed over a sizable area in the interior of southern Africa. In his well known *Worlds in Collision*, Immanuel Velikovsky gets around the lack of African flood traditions with an ingenuity that must be admired: he invokes a "collective amnesia" [253:302] which very conveniently blotted the disaster out of the memory of certain whole societies, just as completely as terrible events may be erased from the memory of an individual.

Turning to the other side of the world, Australia has several different flood traditions which have little in common with each other and nothing in common with the Bible story; these could be independent recollections of local floods. According to the aborigines in the Lake Tyres area of Victoria, a gigantic frog once swallowed all the water in the world. The animals tried to make him laugh and so disgorge the water, but none of their antics brought so much as a smile until the eel stood on its tail and danced. At this the frog laughed so heartily that the water gushed out in a terrible flood and many were drowned [197:175]. In another legend from Victoria (perhaps part of the same one?) the survivors of a great flood owed their lives to the pelican, who picked them up in his bark canoe [2:56].

From Western Australia comes a story with a rather obvious moral: Long ago two races lived on opposite shores of a great river, one white and one black. They intermarried, feasted together, and fought each other in a friendly fashion. The whites were more powerful and had better spears and boomerangs, so they came to feel superior and broke off relations with the blacks. This situation existed for some time, until one day it began to rain. For months it poured, and the river overflowed and forced the blacks to retreat into the hinterland. Eventually the rains ceased and the waters receded. When the blacks returned to their old hunting grounds they were astonished to find that their proud neighbors had vanished completely under the waters of a wide sea. According to Andree [2:33-34] this legend is old, but the detail that the proud race was white was added after the colonization of Australia by the British.

The South Seas also provide abundant flood traditions in very diverse forms. Aside from some biblical parallels, which can easily be attributed to missionary influence, many of these traditions are remarkably consistent with the local geologic setting. Very often the flood is said to have come from the sea, as would be expected in islands frequently subjected to earthquake-generated tsunamis or typhoon-lashed waves.

In the Society Islands there is a legend associated with the island of Raiatea [2:39-40]. One day a fisherman, through either ignorance or just plain disobedience, violated a tabu by fishing in waters sacred to the sea god Ruahatu. His hook caught the sleeping god by the hair, and after a long hard struggle he hauled up a very furious deity indeed. Berating the fisherman, Ruahatu decreed that the land was now defiled and must be destroyed. The fisherman threw himself down and begged for mercy, imploring the god at least to let him escape. Ruahatu relented and ordered the fisherman to betake himself with his family to the islet of Taomorama, inside the reefs on the east side of Raiatea. Next morning the waters of the ocean began to rise and the people on Raiatea took to the hills. Finally even the tops of the peaks were swamped and everyone was drowned. When the waters receded the fisherman and his family returned to the main island and became the progenitors of the present inhabitants. This legend is considered by Andree to be an attempt to account for fossil corals and shells found above sea level, dating from former high stands of the sea relative to the land. The Raiateans do not seem to be troubled by one very glaring inconsistency in the tale: the highest point on Raiatea is 3,388 feet above sea level, while Taomorama, the place of refuge, is a tiny coral islet only a couple of feet above high tide level at its highest point! Nothing could be further from the biblical tradition than that the sinner is saved while the innocent are destroyed—the same Polynesian attitude as that reflected in the legend of Kahawali, which so disturbed William Ellis (see Chapter 6).

The very same attitude is reflected in a flood tradition from the Fiji Islands [2:37-38]. In this case the flood (in the form of unceasing rain) was sent by the great god Ndengei, to punish his wicked and unrepentant nephews for killing the Turukawa bird. As the waters rose toward the high peaks where they had taken refuge, the culprits were saved by one of the lesser gods to whom they appealed. Some versions say that they were advised to build a raft out of the fruits of the pompelmous, others that two canoes were sent, and still others that they were taught to build a dugout. In any case, they floated to the island of Mbengga, whose inhabitants consider themselves the Fiji equivalent of our Mayflower descendants.

The decidedly unusual topography of Mangaia, one of the Cook Islands, is clearly reflected in the local flood tradition. A flat-topped central core of eroded volcanic rock, 554 feet above sea level at its highest point, is surrounded by a sort of moat only 20 to 40 feet above sea level, in which taro is cultivated. This in turn is ringed by a raised platform of eroded coral rock called the *Makatea*, 110 to 210 feet above the sea. From the base of the Makatea at an elevation of about 45 feet, a 100-foot-wide terrace slopes gently toward the sea, ending in a cliff 15 to 35 feet high. At sea level, living coral forms a fringing reef 300 yards wide. Mangaia's shape is unique (Fig. 29).

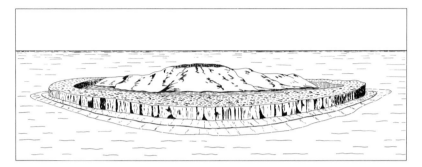

FIGURE 29. *The island of Mangaia in the South Pacific, whose unique shape inspired a local flood tradition. The center of the island is an ancient volcano which was worn down to sea level, then uplifted and eroded, while a fringing reef grew around it. Gradual subsidence permitted the coral to grow upward to form a barrier reef separated from the island by a lagoon. Re-elevation left the barrier reef high and dry, forming the "Makatea," a platform surrounding the moat-like depression which was the lagoon. (After Marshall, 1927 [153].)*

The local myth relates that the shape of the island was once smooth and regular, with gentle slopes. One day the gods of the sea and of the rain decided to engage in a contest to see which was the more powerful. With the help of the wind god, the sea god attacked the island and succeeded in inundating the coast to the height of the Makatea. Then the rain god caused it to rain for five days and nights, washing the red clay and small stones into the ocean and carving the deep valleys in the slopes, until only the flat top of the central peak remained of the original surface. Having been warned of the impending struggle, the first chief, Rangi by name, had led his people to this central peak—the "Crown of Mangaia." As their situation became increasingly precarious, Rangi appealed to the supreme god, who ordered the others to call off their contest.

This folklore explanation of Mangaia's shape reveals a definite, though misapplied, appreciation of the role of running water and of sea waves, particularly storm-driven waves, in shaping the landscape. The geologic history of the island, in brief, is roughly this [153]: Originally the island was a volcano built up from the sea floor. After all activity had ceased, it was eroded by the combined attack of running water and waves until nothing remained but a shoal. This later was elevated above sea level and subjected to prolonged weathering (decomposing the rock surface to red clay) and erosion by running water (carving the slopes into deep valleys). In the meantime a fringing reef grew around the edge of the island. Then the island subsided so gradually that the coral was able to grow upward to become a barrier reef, separated from the land by a lagoon; in the later stages of subsidence it also grew outward. Reelevation of the land left the barrier reef high and dry, forming the Makatea; the seaward extension of the barrier reef became the terrace slope, and the former lagoon behind the reef became the moat-like taro flat. In the latest stage of development a new fringing reef has grown outward from the edge of the terrace. Frazer remarks:

> Had the writer who records the tale not only described the aspect of the island . . . we should probably have failed to perceive the purely local origin of the story, and might have been tempted to derive it from some distant source, perhaps even to find in it a confused reminiscence of Noah and the ark. It is allowable to conjecture that many other stories of a great flood could similarly be resolved into merely local myths, if we were better acquainted with those natural features of the country which the tales were invented to explain.
> [63:248-249]

North American Indian lore contains abundant evidence of the way in which primitive mythologies absorb later elements. For instance, Old Coyote Man, the culture-hero of the Crow legends, is supposed to have invented horses—but horses were unknown to the Indians until the Conquistadores introduced them in the sixteenth century. The general resemblance of many of the American Indian flood traditions to each other can readily be explained in terms of migration or contacts between tribes, and frequent resemblances to the Bible story are not at all difficult to attribute to the efforts of missionaries.

A Chippewa legend [2:82] is unusual in that it attributes the flood to the melting of snow. In the beginning of time, in the month of September, there was a great snow. A little mouse nibbled a hole in the leather bag which contained the sun's heat, and the heat poured out over the earth and melted all the snow in an instant. The meltwater rose to the tops of the highest pines and kept on rising until even the highest mountains were submerged. One old man had foreseen the flood and warned his fellows, but they had just replied, "When it comes, we'll just take to the hills." They were all drowned; but the old man had built a large canoe in which he drifted on the flood, rescuing whatever animals he encountered. After a while he sent out the beaver, the otter, the muskrat, and the duck in turn, to try to find land. Only the last returned, bringing mud in its bill. The old man cast the mud on the waters and blew on it, and it expanded into an island large enough to hold him and all the animals.

Even in the relatively arid southwest there are flood traditions. According to the Papagos [63:281-282], the Great Spirit first created the earth and all living things except man. Next he made the hero Montezuma, with whose help he made all the Indian tribes. The first world was a happy and peaceful one, but it was destroyed in a great flood. Warned by his friend Coyote, Montezuma had built a boat and when the waters rose the two of them were saved. After the waters receded, the Great Spirit, with Montezuma's help again, created men and animals anew.

A flood story from Arizona belongs in the category of "ex post facto" geomyths, made up to account for a detail of the landscape. It tells how, as the waters rose, a great chief led his warriors higher and higher into the Superstition Mountains. When it became apparent that even the highest peak would be submerged, the chief turned his braves to stone rather than let them drown ignominiously, and there they stand today, guarding the heights (Plate 35) [215]. The Apache Warriors in fact are weathered-out columns of a Miocene welded tuff.* I do not know whether this is an authentic legend or not, but the Superstition Mountains, some twenty miles east of Phoenix, are held sacred by the Apaches. They believe that the hole leading down into Mother Earth is located in this range, its

* Pumiceous rock in which the particles have been fritted together by the intense heat of burning eruptive gases. Welded tuffs closely resemble lava flows after they cool. Columnar jointing is often well developed in them.

PLATE 35. *The "Apache Warriors" on the Superstition Mountains, Arizona. (Photo by Michael F. Sheridan.)*

entrance guarded by a nine-headed snake which allows no mortal to pass. Winds blowing out of this hole are supposed to be responsible for severe dust storms. The emphasis in the tale of the Apache Warriors is not on the flood itself, but on the creation of a dramatic landform which fairly cries out for explanation. If the tale is not modern "fakelore" and the flood is said to be *the* deluge, I would be inclined to suspect a merging of traditions after the Apaches learned of Noah's flood.

Incidentally, I am told [162] that there are Americans today who point to a light-colored band of rock high in the Goldfield Mountains of Arizona as the high-water mark of Noah's flood (Plate 36). The light band is a rhyolite tuff. Underlain by dark gray Precambrian granites and overlain by a dark lava flow, the tuff stands out so prominently that on a clear day it can be seen from Phoenix, twenty-five miles away.

In a legend of the Makah Indians of Cape Flattery, Washington [2:91-92], the Pacific Ocean once rose and fell several times in the course of a few days. First the water rose high enough to cut off the cape from the mainland, then it suddenly withdrew leaving Neah Bay high and dry. Four days later it reached its lowest level, and then it welled up again quietly until the cape and all the land

PLATE 36. *The Goldfield Mountains of Arizona, showing the light band of rhyolite tuff believed by some people, even today, to be the high-water mark of Noah's flood. Looking northeast from the Bush Highway. (Photo by Michael F. Sheridan.)*

were submerged except for the mountain tops. The rising water was very warm. Those who had canoes loaded their belongings and were borne hither and thither, but generally northward, on a strong current. Some of the canoes were caught in trees and many people lost their lives. When the sea returned to normal after four more days, part of the tribe found itself far to the north, where their descendants still dwell.

J. G. Swan, who according to Andree first reported this legend in 1869, attributed it to some "volcanic rise and fall of the land." However, the immediate area is not volcanic, and in any case such upheavals are not known in connection with volcanism. How much simpler to say that it was the sea level itself that rose and fell, as the legend specifies. Allowing for some exaggeration, no better picture could be painted of a tsunami resulting from some distant earthquake.

Flood traditions are prolific throughout Latin America. There are numerous legends in which the survivors of the deluge, either a couple or a family, escape in a calabash, a canoe, or a raft, or climb mountains or trees. Biblical overtones are very recognizable in some cases.

A prime example of a flood legend of purely local and independent origin is that of the Araucanian Indians [2:117], whose home was the part of Chile once known as Araucania (Fig. 30). Two great serpents are said to have made the sea rise just to prove which had the more powerful magic. The flood came after a strong earthquake, connected with a volcanic eruption, and the people took refuge on a mountain which floated up close to the sun. Thereafter whenever the Araucanians felt an earthquake they took to the hills, carrying bowls to protect their heads from the sun's heat. Along that part of the Pacific coast there are active faults, movement on which caused the earthquakes of May 1960 [53]. The tsunami generated by that shock proved disastrous for many. Earthquakes accompanied by tsunamis must have been frequent throughout the ages. Also significant, possibly, is that one of the three known instances where a volcanic eruption seems to have been triggered *directly* by an earthquake is that of Puyehue, located on one of the faults on which movement occurred; it went into mild eruption within two days after the first shock [111], probably because the earthquake allowed ground water to come in contact with hot magma (as in another known case, that of Pematang Bata on Sumatra in 1933). Need any other source be sought for the Araucanian flood tradition than some similar chain of circumstances in the past? That the disaster was attributed only to the caprice of supernatural creatures, rather than to punishment for some transgression, speaks well for the clear conscience of the Indians who were the original sufferers.

Bogota, the capital of Colombia, lies on the Cundinamarca plateau surrounded by mountains. The plateau is drained by the Funza River, which flows through a narrow gorge in the mountains bounding the plateau on the west and drops over the Tequendama Falls, about 450 feet high, on its way to join the Magdalena River. The flood tradition of the Chibcha Indians [89:448], like the traditions accounting for the Vale of Tempe and the Vale of Kashmir, is an imaginative attempt to explain the local geography: Long ago the people who lived on the plateau were wild savages. The sun god Bochica came among them as an old man and taught them how to build huts and organize themselves into an orderly society. His beautiful but wicked wife, Chia, tried to frustrate his efforts to civilize the Indians, but her magic was not as powerful as his. The

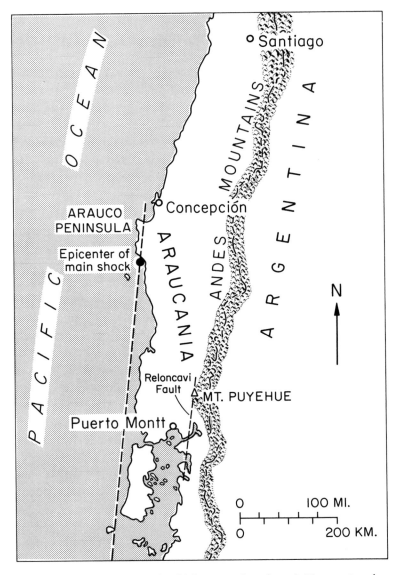

FIGURE 30. *The region of the Chilean earthquakes of May, 1960. A tsunami like the one generated by the main shock could have been the source of the Araucanian Indian flood tradition.*

best she could do in the way of mischief was to cause the Funza to overflow and flood the whole plateau, which at that time was surrounded by an unbroken wall of mountains. Only a few people, who reached the mountains in time, escaped drowning. Bochica in anger banished Chia to the sky, where she became the Moon. Then he rent the mountain wall and the floodwaters drained away, leaving only Lake Guatavita.

In the Rio de Janiero region of Brazil the Indians had a flood legend in which two couples were saved by climbing tall trees [89:454]. The twin sons of a great wizard were always quarreling, for one was good and the other bad. One day the good brother, angered by some act of the bad twin, stamped so hard that the earth opened and a fountain of water gushed up as high as the clouds. The whole world was soon submerged. The good brother and his wife climbed a *pindona* tree, the evil brother and his wife a *geniper* tree, and there they clung until the waters receded. From these two couples descended the Tupinambas and Tominus, two tribes that were always at odds with one another. Any local flood could have been the origin of this legend.

As is to be expected from geographically contiguous peoples, the Aztecs, Mayas, and Quichés of Mexico and Guatemala had flood traditions with many elements in common. In all of them the flood is part of the creation myth, in which there are several attempts to make man, with the unsatisfactory results of early attempts being destroyed in a series of cataclysms. As told in the *Popul Vuh* of the Quichés [2:109-111], the gods were not satisfied after they had made the animals, for the animals neither spoke nor honored their creators; so they tried to make man out of clay. The first people could not turn their heads, and though they could speak, they understood nothing. Therefore the gods sent a flood which destroyed their defective handiwork. In a second attempt they fashioned man out of wood and woman out of reed, but though these people were better than the first they were still animal-like, spoke unintelligibly, and were not grateful to the gods. Most of these were destroyed by a rain of burning resin and an earthquake, but a few survivors ran off into the woods and became monkeys. The third time the gods used white and yellow corn, and the results were so good that the gods were alarmed and took back some of the superhuman characteristics with which they had endowed man, leaving the imperfect kind of human beings we have in the world today to become the ancestors of the Quichés.

In the Aztec account the first race of man was devoured by jaguars (or ocelots), the second swept away by wind and turned into monkeys, the third showered by fire (an eruption?) and turned into birds, and the fourth overwhelmed by a flood and turned into fish, except for one couple who were saved in a hollowed-out cypress log. The most common Mexican version has it that the man who escaped was named Coxcox and his wife, Xochiquetzal; in other versions they are a god and goddess. The little bark in which they escaped was left stranded on Culhuacan peak. Many children were born to them, all mute. Then along came a dove, which gave them tongues and endowed them with many different languages, and from these children descended all the different nations of the world.

The story of Coxcox is the one and only flood legend with possibly biblical elements for which there seems to be pre-missionary documentation in the form of pictographs. Or is there? According to Andree [2:105] none of the early writers concerned with Mexican mythology, who could have heard the tale at the time of the Conquistadores or shortly after, ever mentioned a Bible-like flood legend, and he doubted that the interpretation of the pictographs was the correct one. In this he followed Don Jose Fernando Ramirez, Conservator of the National Museum in Mexico City, who showed that the descriptions of the pictographs as given by Clavigero, Humboldt, Kingsborough, and others were all based on the same source, a picture map published by Gemilli Careri in Churchill's *A Collection of Voyages and Travels*, volume 4 [29]. Gemilli Careri had read into this picture the story of the Flood, and Humboldt and all the rest followed suit and accepted his interpretation. But according to Ramirez the "dove" was intended to be the bird known as the Tihuitochan, which calls "Ti-hui," and the picture actually represented the story of the migration of the Aztecs to the Valley of Mexico. The Aztecs are believed to have come into Mexico from farther to the north. Their traditions told how a little bird kept repeating "Ti-hui, ti-hui," which in their language meant "Let's go!" and their priests interpreted this as a divine command to seek a new home. Seven subtribes set out, six of whom established themselves more or less quickly in various parts of Mexico, while the seventh wandered for some time, looking for a sign in the form of an eagle sitting on a rock holding a serpent in its mouth. The promised sign was encountered at Lake Texcoco, and accordingly the city now known as Mexico City was founded on its shores in 1325. This, then, is the tradition historians believe is embodied in the picture writing in question; it was Gemilli Careri alone who decided that the bird in the picture was the dove giving out tongues. He himself admitted that the chronology was "not so exact as it should be, there being too few years allow'd between the flood and the founding of Mexico" [29:485]—for the picture includes symbols telling the number of years spent in various places during the wanderings.

Gemilli Careri heard the story of Coxcox during his sojourn in Mexico in 1667, well over a hundred years after the first missionaries had arrived with Cortez and ample time for biblical details to have become superposed on indigenous Aztec myths and traditions. Other Mexican flood stories are quite obviously the Bible story transplanted to a more familiar local setting. In Michoacan, for instance, the central character is named Tezpi instead of Noah, but other than that the main departure from the biblical version is that Tezpi first sends out a vulture to find land, which finds so many corpses that it does not return; then he sends out other birds, and finally the hummingbird returns with a green twig in its bill [63:275-276].

To cite further examples of flood traditions would become tedious, if it has not done so already. Enough instances have been given, I hope, to demonstrate that when viewed from their geologic context, many flood traditions obviously have originated on the spot. I can see no reason to assume that in explaining the ubiquity of flood traditions we are limited to a choice between two extreme alternatives. Velikovsky, for instance, states:

The answer to the problem of the similarity of the motifs in the folklore of various peoples is, in my view, as follows: A great many ideas reflect real historical content. There is a legend, found all over the world, that a deluge swept the earth and covered hills and even mountains. We have a poor opinion of the mental abilities of our ancestors if we think that merely an extraordinary overflow of the Euphrates so impressed the nomads of the desert that they thought the entire world was flooded, and that the legend so born wandered from people to people. [253:308]

To which one might reply: Of course many ideas reflect real historic content. However, there is not *one* deluge legend, but rather a collection of traditions which are so diverse that they can be explained neither by one general catastrophe alone, nor by the dissemination of one local tradition alone. Some are highly imaginative but very wide-of-the-mark attempts to explain local topographic features or the presence of fossil shells high above sea level. A large number are recollections—vastly distorted and exaggerated, as is the rule in folklore—of real local disasters, often demonstrably consistent with special local geologic conditions. Surely it is not accidental, for instance, that in many flood traditions from the Pacific coast of the Americas and from Pacific islands, the flood is attributed to a rise of the sea; more than 90 percent of the earthquake energy released annually in the world is released in the Pacific area, and consequently tsunamis are most likely to be generated there. One of the oldest of the remembered flood disasters occurred a long, long time ago in Mesopotamia, and it made such an impression on the dwellers in the city of Ur that the tale was handed down from generation to generation and carried with the Patriarchs when they migrated toward the Mediterranean. The legend born of that long-ago flood might never have wandered very much farther from its source were it not for the fact that it became a part of the Scriptures, and thus in later ages was zealously carried to every corner of the world by Christian missionaries, often to become merged with pre-existing traditions indigenous to their localities. Flood traditions are nearly universal, partly because of the efforts of these missionaries, but mainly because floods *in the plural* are the most nearly universal of all geologic catastrophes.

8 /

The Minoan Eruption

of Santorin

IN CHAPTER 2 WE GAVE EXAMPLES TO ILLUSTRATE how volcanic eruptions have affected the destiny of whole communities and even of whole nations, albeit small ones. What would be the result if an eruption of unprecedented violence occurred in the very heart of the civilized world? There was a time when the Mediterranean area was the hub of European, Egyptian, and Near Eastern civilizations, and during that time there was an eruption whose violence possibly has never been equalled in the memory of man anywhere on earth—the Bronze Age eruption of Santorin, whose possible impact on history and legend has become the subject of numerous speculations. In recent years it has been proposed that this eruption might have been responsible, directly or indirectly, not only for Deukalion's deluge [73] but also for the sudden collapse of Minoan civilization on Crete and the concomitant rise of the Mycenean civilization on the Greek mainland [148]; for the myth of Atlantis [73]; for the plagues of Egypt [12] and the miracle of the parting of the Red Sea waters [67]; for the myths of Phaëthon [74] and Icarus [22], parts of the Theseus myth [200, 201], and details of the myths of Amphitryon [156:252], the Argonauts [143], and Talos [214]; and for numerous semihistorical traditions in the Aegean and eastern Mediterranean, particularly of local inundations [143]. Before we can judge the validity of such

179

sweeping claims, it is obvious that we must take a very close look at the volcano and at the circumstances of its Bronze Age eruption.

Santorin volcano comprises five islands which constitute the southernmost unit of the Cyclades Islands in the Aegean (see Fig. 28). The highest point, the Prophet Elias massif on Thera, represents the original island made of schist and marble. Some time in the late Pliocene period a volcano was born, probably just off the west shore, and it grew in successive eruptions from several vents until a large edifice was built up, covering most of the older rocks of the island. To its Bronze Age inhabitants the island was known as Stronghyli ("Round"). In the middle of the fifteenth millennium B.C. there was a tremendous pumice eruption, similar to that of Mount Mazama in Oregon a few thousand years earlier and likewise culminating in caldera collapse. Covered with a thick layer of tephra, the remnants of Stronghyli remained uninhabited for an unknown length of time, possibly a couple of centuries at least. Then the Phoenicians seem to have planted a colony on the largest remnant [46], which they renamed Kalliste ("Most Beautiful").* Their leader is said to have been the legendary Cadmus, who rediscovered the island in his search for his sister Europa after she had been carried off by Zeus in the guise of a bull. In its subsequent history the little cluster of islands changed hands many times. Only two of its rulers warrant mention here, because they account for the two current names. In the ninth century B.C. the group was conquered by Spartans under a leader named Theras, from whom we get the name Thera, which is used to designate the largest island alone and also the whole group as a political unit; and under Venetian rule the group became known as Santorini (of which Santorin is the frequently used French form), which is a corruption of the medieval Italian name for Saint Irene, the patron saint of Thera [46]. Today the names Santorin(i) and Thera are usually used interchangeably. In these pages the name *Thera* will be re-

---

* According to some, Kalliste was one of the pre-eruption names for Santorin, and surely it would have been more appropriate then than afterward. If the name does date from after the eruption, it either referred to the dramatic beauty of the sheer and colorful cliffs encircling the bay, or was applied sardonically to the desolate landscape above, in the same way that Robin Hood's large lieutenant was known as "Little John," for example.

served exclusively for the main island, and *Santorin* for the volcano
—which comprises the group of islands as a whole.

Immediately after the caldera collapse, Santorin consisted only
of Thera, Therasia, and tiny Aspronisi islands (Fig. 31). But the
volcano was not asleep, and it began unobtrusively to build a new
edifice within the ruins of the old. The first historical record of any
activity dates from 197 B.C.; at that time a new island poked its
head above water in the middle of the bay, and was given the appropriate name of Kameni ("Burnt"). Again and again eruptions added
to the growing underwater mass. There are reports, not all of the
earlier ones verified, of fourteen eruptions between A.D. 19 and 1950
[80]. In 1570 a second island appeared, and in the 1707-11 eruption
another came into being between the first two. In 1866 a new dome
emerged above sea level and finally, in the last major eruption, that
of 1925-26, lava flows united all but the oldest island, Palea ("Old")
Kameni, to form the present Nea ("New") Kameni.

The inhabitants of Santorin today live in an impressive but
precarious setting. The precipitous cliffs of crescent-shaped Thera
fall off sharply from heights of 500 to 1,200 feet into water so
deep that ships cannot anchor below the main town of Phira, but
must tie up to buoys fastened to the sea bottom by extra-long
chains, or anchor close to Nea Kameni; in either case they must send
passengers and goods ashore by launch. A steep path winds up the
cliff face from the main landing place to Phira, perched on the rim.
No vehicles can negotiate that sharp zigzag turn; visitors landing
there must make the ascent by donkey or by shank's mare. It would
be hard to say which is the more breathtaking, the view of Thera by
night or by day. My first sight of it was at night. The steamer had
arrived while we were at dinner, and by the time we came on deck
it was fully dark, so that the whole spectacle burst upon the sight
with full impact. A cluster of bright lights marked the landing at
the water's edge; a crown of brilliants marked Phira, strung out
along the cliff's edge high above; and a jagged line of lights marked
the donkey path snaking its way between the two.

By day the cliff face can be seen to be made up of layers of
multicolored ash and lava. All shades are visible, from brick red
and deep rose to palest pink, from darkest brown to pale buff, and
from black to white, all contrasted against the strong indigo blue

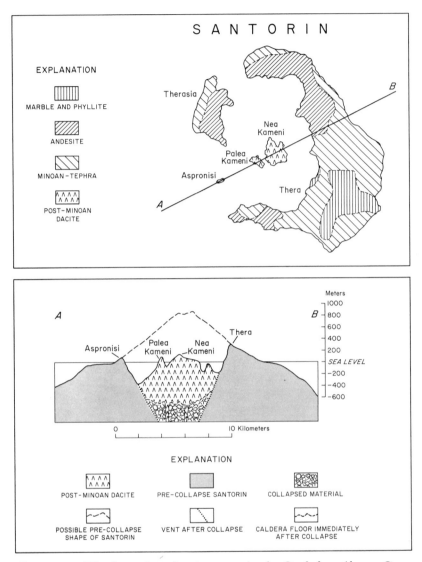

FIGURE 31. *The Santorin volcanic group in the Cyclades. Above: Geologic map of the group. Below: Section along the line AB. The conjectured depth of the caldera floor immediately after collapse may be somewhat exaggerated.*

PLATE 37. *The wall of the Santorin caldera as seen from the bay. Looking eastward. The vertical wall of pumice, which once reached to the very edge of the cliff, has been pushed back by quarrying until hardly visible from this vantage point. The light-colored slopes at the foot of the cliff are pumice which has been mined out and dumped over the edge to be loaded on ships.*

of the sea below and the equally startling azure of the Mediterranean sky above. The whitewashed houses of Phira and of the villages at other points on the rim add a dazzling accent. Near the landing place below Phira there is a steep, grayish-white artificial talus* cone, formed of pumice dumped over the cliff from the quarries above, waiting to be loaded into ships (Plate 37). From certain points in the bay, the once-smooth skyline directly above this talus presents a gap-toothed appearance, where blocks of pumice have been mined out leaving vertical walls. The outer slope of Thera is gentle, falling off gradually to a smooth sandy coast which affords no good protected harbors. Most of the people on Santorin live in Phira and several large villages, including one on Therasia.

Life is precarious for more than one reason. Much more to be feared than the recurrent eruptions are the earthquakes which rock the area from time to time. After a severe one in 1956, which killed dozens of people, many moved away [257]. Those who remained support themselves, and not too well in most cases, by fishing, agri-

* Talus, or scree, is a heap of fallen disintegrated material forming a slope at the foot of a steeper declivity.

culture, and mining. Rainfall is so low that water for domestic purposes is brought in once a week in a huge plastic container towed behind a ship. The highly porous pumiceous soil absorbs and holds what little rain does fall, which is why cultivation is possible at all; summers are hot and dry. Everything that grows, even the vines, must huddle close to the ground in shallow depressions for protection from the strong winds which are prevalent. The pumiceous ash which is mined there is of the type called *pozzuolana*, used to make hydraulic cement—that is, cement which sets under water; the Santorin pozzuolana, which was used in building the Suez Canal, is the tephra produced in the Bronze Age eruption.

Santorin, a superb example of a caldera formed on the sea floor, is a classic area for volcanologists. The island of Thera is also important for archeologists, for the new excavations there promise to shed light on crucial questions concerning Minoan archeology [257]. Since the question of Santorin's possible connection with Atlantis is linked (though not inextricably) with the question of its possible connection with the collapse of Minoan civilization, we will have to leave Atlantis alone for the time being until we have established first, what happened or could have happened and, just as important, what could not possibly have happened, as a result of the Bronze Age eruption of Santorin; and second, how whatever did happen or could have happened would have affected the Minoans on Crete.

A caldera even better known than that of Santorin is that of Krakatau (Krakatoa) in the Sunda Strait between Java and Sumatra (Fig. 32), formed in the eruption of 1883. As we have quite detailed and reliable accounts of that eruption and its consequences [66, 254], we can draw many valid inferences concerning the Santorin eruption in Minoan times, for it was of the same type. Before the 1883 eruption, Krakatau consisted of three coalescing cones in a row. In prehistoric times it had been a single cone, perhaps more than a mile high, which collapsed after a violent eruption to form a caldera, completely under water except for some small islands on its rim. The three new cones had been built up in subsequent eruptions, which had begun to form a new volcanic edifice within the prehistoric caldera just as the Kameni islands are building up today within the present Santorin caldera. The 1883

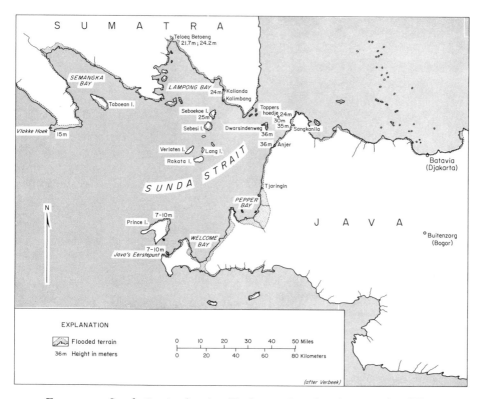

FIGURE 32. *Sunda Strait, showing Krakatau after the 1883 eruption (Verlaten, Rakata, and Lang Islands comprise its remains) and the extent of inundation by the main tsunami. Heights reached by the wave were generally greater toward the narrow eastern end of the strait, at the heads of bays, and on the sides of islands facing the volcano. On tiny Toppershoedje, right in the mouth of the strait, the wave was nearly twenty feet higher on the southwestern side than on the northeastern.*

activity began, mildly enough, on May 20. The explosions were not particularly alarming and soon subsided. Activity resumed on June 19, and by August 11 all three cones were in a state of mild explosive activity. The first serious explosion occurred at 1:00 P.M. on August 26. Explosions of increasing severity continued until 5:00 P.M., when the first collapse occurred, and continued throughout the night, keeping people awake as far away as Batavia (now Djakarta) and Buitenzorg (now Bogor). About 10:00 A.M. on the twenty-seventh the eruption reached its grand climax, during which the ash cloud reached a height of fifty miles and the main collapse occurred. Explosions continued with diminishing intensity throughout the remainder of the twenty-seventh and the early morning of the twenty-eighth, and then it was all over [25:80-83].

As a consequence of the enormous amounts of ash that were blown into the air, areas as much as 275 miles away were plunged into total darkness; 130 miles away the blackout lasted twenty-four hours, and 50 miles away, fifty-seven hours [25:83]. In the immediate vicinity of Krakatau it lasted three days and was so thick that nobody could see his hand before his face, literally; lamplight hardly penetrated the gloom. Dust fell on ships 1,600 miles away three days afterward, and fine dust remained suspended in the high atmosphere for years, producing spectacular sunsets all over the world. The sea in the neighborhood of the volcano was littered with a thick layer of pumice.* Ships that plowed their way through the floating pumice reported that it was ten feet thick in places. An iron bar thrown onto it from one ship did not sink, and tree trunks and corpses were seen embedded in it [66:154]. Fist-sized lumps of pumice were hurled twenty-four miles from Krakatau, fine lapilli twice as far [254:127]. Aerial vibrations took the form of sound waves or shock waves, depending on their wave length. The roar of the explosion was heard at Rodriguez Island in the Indian Ocean, nearly three thousand miles away. Air pressure waves blew out gas burners, upset lamps, broke shop windows 80 miles away, and cracked the walls of buildings as much as 480 miles away [254:363].

---

* Pumice, a typical product of the very highly explosive kind of eruptions with which many calderas are associated, is formed when molten magma "froths" in the vent and bursts its way out violently. All magmas contain quantities of volatile materials—water and gases—which remain dissolved in the liquid phase so long as the pressure is high enough. As the magma nears the surface, the pressure decreases and the gases and steam are released, just as the bubbles in champagne are released when the cork is popped. The volatile content of the magma, and the suddenness of its release—or in other words, the explosivity of the eruption—depend on several factors ultimately related to the chemical composition of the magma, its temperature, and the pressure. Magmas of basaltic composition (low in silica) are more apt to flow out in liquid form as lava, and are seldom dangerous to anybody or anything not right in the path of the flow. On the other hand, magmas of andesitic to rhyolitic composition (containing intermediate to high amounts of silica) are more apt to explode violently, in which case the lava in the vent is shattered into ash or froths into pumice and is ejected in the form of tephra. Pumice is so light and full of air that it can float long distances before it becomes waterlogged and sinks, or before it is washed ashore many miles from its source.

Practically all the loss of life in the Krakatau disaster was caused by the tsunami generated by the main collapse. That wave destroyed 295 villages and drowned at least 36,000 people on the nearby coasts of Java and Sumatra. It reached its maximum height of 36 meters (about 120 feet) at two places, at Anjer on the Java coast and on the south-facing side of the little island of Dwars in den Weg in the narrowest part of Sunda Strait (see Fig. 32), both over thirty miles from the source; it was only about 82.5 feet high on Seboekoe Island, half the distance from Krakatau but protected behind Sebesi Island; and on the tiny islet of Toppershoedje, also in the narrowest part of Sunda Strait, there was the substantial difference of nearly 20 feet in the height attained by the wave on the side facing Krakatau and that attained on the lee side [254:408].

Only a few very mild earthquakes were perceived in connection with the Krakatau eruption, and those only in the immediate vicinity. Early reports of earthquakes in more distant parts of the Indonesian Archipelago are believed to have been based on the effects of the aerial shock waves rather than of actual ground movements [254:123]. As is the case in shallow shocks, particularly volcanic earthquakes, the seismic energy generated by the Krakatau explosions must have been dissipated rapidly away from the source. (Of course, had sensitive seismographs been known at the time [47], the seismic waves generated by the Krakatau explosions would have been recorded around the world even though not otherwise perceptible.*) While there were no appreciable earthquakes directly associated with the Krakatau eruption, however, there seems to have been an increase in the seismicity of the general area afterward; shocks were felt in the Bantam region on September 1 and 18 and December 6, 1883, and in January and February 1884 [254:122].

How violent was the Bronze Age eruption of Santorin compared to that of Krakatau? That is what we must know in order to

---

* Underground nuclear tests are monitored by seismograph networks stretching hundreds to thousands of miles away, to yield information of scientific as well as practical value. Seismic waves from the first of such tests in Nevada, an explosion equivalent to 1.7 kilotons of TNT, were recorded by instruments 370 miles away [90]—and the energy of the main Krakatau explosion is estimated to have been equivalent to 100 to 150 megatons [191], which is roughly 60,000 to 90,000 times greater!

assess its possible effects on Crete and elsewhere. It has sometimes been stated that because the Santorin caldera is four to fives times larger than that of Krakatau, its eruption must have been four to five times more powerful, but that is very much of an oversimplification. What is really relevant is not how much *total* energy was expended, but *how that energy was partitioned* [265]; obviously, the effects of a few tremendous explosions* would be more far-reaching than the effects of numerous smaller ones, even though the total energy would be the same. Similarly, what is relevant in connection with possible tsunami damage on distant shores is not how much of the island collapsed, but how much collapsed *at any one time.* Also important to Minoan archeology are the answers to these questions: How long did the eruption last from start to finish? Exactly when did the collapse (or collapses) occur relative to the climax of the eruption? Were any earthquakes associated with the eruption and if so, how far away might they have been felt? These questions cannot be answered unequivocally from the field evidence.

Three layers of tephra have been recognized on Santorin as the products of the Bronze Age eruption. The lowest is 10 feet thick in places, the second from 17 to 33 feet thick, and the uppermost from 33 to 100 feet thick. One of these layers, apparently the uppermost, can be traced over great distances in deep-sea cores† (Fig. 33) [170]. It is estimated that it covered an area of about 77,000 square miles, and that the cloud of gases, vapors, and dust from which it fell out actually must have covered a significantly greater area [170]. Krakatau's ash, whose atmospheric extent and effects are

---

* Unfortunately, in many works on this subject the words "explosion" and "eruption" have been used interchangeably, but they are anything but synonymous. An eruption does not consist of just one big bang; there are numerous individual explosions of variable intensity.

† Oceanographers use an ingenious device called a "piston corer" to take undisturbed samples of the layers of sediment on the sea floor. It is a long metal pipe, about three inches in diameter, with a plastic liner; it is dropped over the side of the research vessel and allowed to fall to the bottom under its own weight. A triggering device which hits bottom just ahead of the pipe releases a piston, which sucks up a column of the soft sediments into the tube as it penetrates them, somewhat as a hypodermic needle draws a blood sample. A valve closes the lower end as the corer is hauled up. The core, in the plastic liner, is then extruded, dried, and carefully sliced open for study. With the best of luck cores up to about a hundred feet long can be obtained, but the average recovery is about fifteen to thirty feet.

FIGURE 33. *Distribution of tephra from the Bronze Age eruption of Santorin, according to the evidence of the deep-sea cores. (After Ninkovich and Heezen, 1965 [170].)*

so well known, shows up little if any in deep-sea cores. Therefore at least one paroxysm of the Santorin eruption must have been substantially more powerful than the greatest of the Krakatau explosions—if so much material was ejected so high into the atmosphere as to be carried so far.

The middle tephra layer shows signs of cross-bedding, which previously were interpreted to mean that it was laid down in a long series of weak to moderate explosions separated by periods of quiescence. However, observations of nuclear explosions have brought to the attention of volcanologists the phenomenon known as "base surge," a characteristic ring-shaped cloud rolling outward from the base of the vertical explosion column. Base surge has since been noted in volcanic eruptions, particularly in those where water enters the volcanic conduit [164]. It transports material of all sizes with tremendous velocity, and close to the eruption center it can erode channels and deposit cross-bedded material like that on Thera. Thus the presumed evidence for erosional intervals during the deposition of the middle pumice could in fact have been produced quite rapidly—even in a matter of days or hours—and at essentially the same time as the upper tephra layer, which would represent material fallen back from greater heights.

The analogy to Krakatau, valid in general terms, cannot be extended to specific details. Thus, because we know that the Krakatau activity began mildly and reached its two-day climax some three months after the initial outbreak, we cannot assume that Santorin did exactly likewise. Every volcano has its own characteristic style, and some known eruptions of the same highly explosive type have run through their cycle very rapidly while others have lasted a long time. The 1835 eruption of Coseguïna in Nicaragua began without warning and was over in a week [206:42]; in the great eruption of Tambora in Indonesia in 1815, on the other hand, the first signs of activity appeared three years before the climax, although the catastrophic outburst itself was over in two days [167]. Eruptions of Hekla in Iceland typically begin with their most violent phase and then taper off gradually [237]. What volcanologists *can* say with some confidence regarding the paroxysmal outburst represented by the uppermost tephra on Santorin is that it should not have been separated very long in time from the violent outburst represented by the lowest layer, certainly not by as much as thirty to fifty years; for it takes thousands of years for pressures to build up to this type of eruption, and once they are released they would tend to exhaust themselves as rapidly as possible. Beyond that, we can *reasonably* assume that the Bronze Age eruption probably began mildly and built to a climax; that the climax probably was quite rapid and undoubtedly very violent; and that the initial signs of activity might have begun months or possibly even a very few years before the final catastrophe.

The distribution of the volcanic ash from the Bronze Age eruption is of vital importance in assessing the effects of the eruption at a distance from Santorin. Several factors are involved in any tephra fall: its thickness, its chemical and physical properties, the time of year, and the climate of the area [236]. The agricultural effects are classed as immediate and long-range; the former are mainly destructive, but over several years or generations the latter may in some cases be beneficial, for even though an ashfall may totally destroy growing crops, "if the deposit is not more than a few inches thick, . . . the next few seasons' plantings may result in normal or even improved harvests because of the beneficial mechanical and chemical effects of the ash that is worked into the old soil" [271]. The thicker the ash or the drier the climate, the slower the recovery. Pasture grass and low-growing foliage are most apt to be smothered, naturally; for higher bushes and trees, the main damage is breakage of branches.

At Paricutín, the volcano born in a Mexican cornfield in 1943, corn grew higher and coffee bushes bore larger berries where the ash was less than a foot thick, for the ash acted as a mulch to retain moisture [275]. On the other hand, in Iceland just a few inches of fresh volcanic tephra has been known to ruin land for agricultural purposes for a long time. Farms have had to be abandoned for at least a year when covered by four inches of ash, for up to five years when covered by a six-inch layer, and for decades when buried under eight to twenty inches. After the 1947 eruption of Hekla, pastures had to be abandoned temporarily when the tephra layer reached a thickness of only half an inch; water supplies were contaminated, and in some areas sheep began to sicken and die after eating grass which was only lightly dusted with ash, for it contained large enough amounts of fluorine, clinging to the tiny ash particles, to be poisonous [238:22].

From the evidence of the deep-sea cores cited above, it has been estimated that an average of about four inches of volcanic ash blanketed eastern Crete as a result of the Bronze Age eruption. That any ejecta other than fine airborne ash reached Crete is doubtful. However, *if* one or more of the Santorin blasts was substantially stronger than the strongest of Krakatau, as could have been the case —particularly if the collapse of the caldera allowed sea water to come into contact with hot magma right at the climax of the eruption, adding the violence of water flashing into steam to the violence of magma exploding into pumiceous ash—then conceivably a few smallish volcanic bombs could have landed on the near shores of Crete, some sixty-plus miles away.

Answers to questions concerning the collapse of the Santorin caldera are necessarily vague. The collapse of Krakatau accompanied the major paroxysms (or, possibly, the major paroxysms accompanied the collapse) [254:398]; on the other hand, the caldera of Askja [235] in Iceland formed gradually over a period of about fifteen years following an eruption in 1875, and the caldera of Isla Fernandina, one of the Galapagos Islands, was gradually enlarged by one to two cubic kilometers in twelve days immediately following a brief but violent eruption in 1968 [216]. Early investigators at Santorin, seeing the vertical walls of pumice standing on the brink of the caldera, concluded that the collapse must have occurred well after the last ash layer was deposited—that is, well after the end of the eruption. Loose, fresh-fallen ash, they reasoned, could not hold up vertically, therefore there must have been time enough for the material to have become compacted before the heart of the island was engulfed, leaving the steep cliffs. However, it has been observed in New Zealand, for instance, that a bed of sharp-angled fragments of pumiceous ash—which are nothing more than tiny shards of volcanic glass—can acquire the necessary degree of coherence within hours after it is deposited. Thus the verticality of the walls of pumice is of no significance with regard to the timing of the collapse. The available geologic information cannot tell us whether the collapse occurred rapidly, thus generating one or more serious tsunamis, or whether it collapsed in such easy stages that the waves set up, if any, were not particularly destructive on distant shores; but it is most likely that the collapse occurred, or at least began, at or very shortly after the climax of the eruption. The idea that the collapse did not take place until more than two hundred years after the eruption, and was then sudden and complete [186, 187], is not at all plausible from the geologic point of view. But in any case, a tsunami connected with the Bronze Age eruption would have been generated by the collapse, not by the explosions or shocks from those explosions.

Some utterly unrealistic estimates have been made of the probable height of the Santorin tsunami, based on misconceptions as to the generation and propagation of such waves. One view has assumed that the wave was generated by an explosion in which all the energy of the eruption was released at once, that it rose to a height of several thousand feet at the explosion center, and that it spread in every direction as a mountain of water, smashing everything in front of it, inundating the whole central plain of Crete, and sparing only shepherds high in the mountains [12]. Others have pictured the tsunami as inflicting damage to exactly the same level all around the shores of the Mediterranean, up to the one hundred or two hundred-foot contour for instance [156, 187]; but as we have seen, the level to which the water will rise at any particular place depends more on local factors than on the wave's original height.

Patches of pumice found at the heads of valleys on the island of Anaphi, the highest of them at 825 feet above sea level, have been cited as evidence of the height of the Bronze Age tsunami [150, 151], despite the fact that it is much more likely that such pumice even at a lower level would be a remnant of an airborne cover, rather than pumice floated up to such heights by a wave of incredible proportions—nearly four times higher than the highest tsunami run-up ever recorded, which was 210 feet at the southern tip of Kamchatka in 1737.* In any case the question is moot, for subsequent investigations have shown that the pumice on Anaphi is from a much older eruption, dated by radiocarbon as sixteen to eighteen thousand years old [125, 168].

Another calculation of the initial height of the Santorin tsunami as 210 meters, or nearly 700 feet [78:112], was made on the basis of a layer of postglacial pumice found some sixteen feet above present sea level at Jaffa, just short of Tel-Aviv. Not only do these calculations entirely neglect the effect of run-up, but they are based on the inverse-square-root-of-distance formula (see Chapter 7), which for mathematical reasons works only in one direction (given the initial height and ideal conditions, it can roughly predict the height at a certain distance from the source, but gives grossly exaggerated values when used in reverse, to find the initial height

---

* Higher waves have been recorded, but they were not tsunamis. In discussions of the possible height of the Santorin tsunami, comparisons have sometimes been made with the giant wave in Lituya Bay, Alaska, in July 1948 [161]. On that occasion an earthquake-triggered landslide crashed into the head of a steep, narrow, T-shaped bay. It sent a sheet of water up and over a 1,720-foot promontory directly opposite the slide and also created a wave which traveled seaward as a huge wall of water. The height of this wave could be measured quite accurately because it completely denuded the forested slopes. Its maximum height was 680 feet near the source, and it diminished to about 35 feet near the mouth of the bay. (A similar wave created by a landslide in Lake Loen, Norway, in 1936 was 230 feet high.) But only after that wave left Lituya Bay could it have started to behave as a tsunami, just as any wave generated by the Santorin collapse would have started to behave as a tsunami only after it had left the confines of its new-formed bay. The pertinent question is, how high did it rise on shores far from Santorin?—and the initial height is only one factor entering into that problem.

from the presumed amplitude on a distant coast). Thus even if the pumice in question were the Minoan pumice, the estimate is meaningless—and an analysis of the heavy and light minerals in it [124: 163] has since ruled out Santorin as a possible source!

The actual propagation of any Santorin tsunami, whatever its initial height, must have been extremely complex. Since three islands remained encircling the void created by the collapse, the wave could not at first propagate freely in all directions, as could a tsunami generated in the open sea. Once outside the caldera the wave fronts, especially the parts traveling eastward around the northern and southern extremities of Thera, must have been considerably complicated by interference (sometimes reinforcing and sometimes cancelling out the amplitude) and by loss of energy upon encountering various islands. The only thing which is quite certain is that any wave or waves generated by the collapse did *not* radiate from Santorin in neat concentric circles, and that it (or they) did *not* reach the same height on all shores, not even at points equidistant from the source. Other than that, we can only say that *if* at least a very substantial part of the caldera collapsed suddenly—which is a very reasonable assumption geologically—then the consequences to the north coast of Crete and the east coast of the Peloponnese could have been serious.

As far as the possibility of earthquakes associated with the Bronze Age eruption is concerned we again can be definite only in a negative way. Since the shocks generated by volcanic explosions have very shallow foci—within a very few kilometers of the surface and probably more on the order of a few hundred meters deep—and since shallow shocks, particularly volcanic shocks, are never felt very far from the source, it is not likely that any seismic waves generated by the Santorin eruption could have been felt as far away as Crete, much less have done damage there. Nor is it likely that a tectonic earthquake would have coincided exactly with the eruption. Although the volcanic belts of the earth are closely paralleled by belts of high seismicity (though the converse is not always the case), the connection between volcanism and tectonic earthquakes is not a simple matter of cause and effect. As mentioned in the last chapter, there are only three instances that I know of where a major earthquake appears to have directly triggered an eruption, Puyehue in the Andes in 1960, Pematang Bata on Sumatra in 1933, and Fuego (Colima) in Mexico in 1973. Almost equally rare are cases where eruptions seem to have caused earthquakes strong enough to be considered tectonic. In 1868 a strong shock shook southeastern Hawaii when a new fissure opened during an eruption of Kilauea. A fairly strong earthquake occurred a few hours after the beginning of the eruption of Sakurajima in Japan in January 1914, which did damage near the volcano and was recorded on seismographs in Europe; and just a few months later a similar earthquake was recorded in connection with an eruption of Iwojima, fifty miles to the south of Sakurajima. These are all borderline cases between volcanic and tectonic shocks. Whatever the connection between eruptions and tectonic earthquakes, it must be sought deep in the earth's mantle (see Fig. 22) [16]. Both phenomena are the result of processes taking place there whose precise nature has not yet been elucidated. With the relationship as remote as that, it is not to be wondered at that tectonic earthquakes and eruptions very rarely coincide.

However, the recorded history of Santorin shows that its major eruptions have been preceded or followed by strong earthquakes with foci at intermediate depth (in the earth's mantle) somewhere in the Mediterranean region [70, 71]. The 1925-26 eruption began in August and ended the following January; on July 6, 1925, there was an earthquake of magnitude 6.5 with its focus 120 kilometers deep under the Peloponnese, and on June 26, 1926, there was another with its focus 100 kilometers deep under Rhodes. The latter had a magnitude of about 8.2 and it caused damage and casualties on Crete, particularly in and near Candia (Heraklion), and destroyed whole villages in the Turkish province of Smyrna. It is entirely within the bounds of possibility that a severe shock of this kind could have occurred within a few years or a few months before or after the Bronze Age eruption of Santorin [93:69], and that such a shock could have been felt throughout the whole eastern Mediterranean region including normally earthquake-free Egypt. And from our vantage point, the results of an event which occurred a little before or after the eruption could easily appear to have been simultaneous with it.

In addition to possible earthquakes linked (albeit remotely) to the eruption, more possible tsunamis, and a very definite ash fall of unknown thickness on Crete, some other consequences of the eruption which certainly must have been perceived at a distance from the volcano would have included a more or less extensive blackout after each major explosion, the effects of which would be psychological everywhere and physical in areas where there was appreciable fallout of ash, such as eastern Crete; tremendous shock waves or loud booms after the most powerful explosions, probably felt or heard far beyond the Mediterranean area, with some of the concussions strong enough to damage buildings within a considerable radius; spectacular electrical displays [15] in the ash cloud over the volcano, which of course would not be visible very far away during the heaviest blackout; very heavy rains and severe thunderstorms associated with the ash particles in the atmosphere, which would act as condensation nuclei ("cloud seeding") for water vapor; a noticeable lowering of temperature whenever and wherever the sun was obscured by the ash cloud; and, of a certainty, extraordinarily spectacular flaming sunsets the world over for many months after the eruption. All these may have left their mark on history or legend, or both. Let us turn first to the people closest to the scene, the inhabitants of Stronghyli itself, and of Crete.

Bronze Age Crete was the home of a civilized people who boasted of a highly refined culture [114, 117, 118, 146, 250, 256]. Sir Arthur Evans, whose excavations at Knossos brought this culture to light, named it the Minoan after King Minos, the sea king of the Theseus myth. Thucydides, in the first chapter of his *Peloponnesian Wars*, states :". . . the first person known to us by tradition as having established a navy is Minos. He made himself master of what is now the Hellenic sea, and ruled over the Cyclades, into most of which he sent the first colonies, expelling the Carians and appointing his own sons governors; and thus did his best to put down piracy in those waters, a necessary step to secure the revenues for his own use" [246:5]. Until Evans's discoveries, however, Thucydides' reference to Minos was generally believed to be to a mythical, rather than to an historial, personage.

Evans divided the Minoan epoch into three main periods, Early, Middle, and Late, each in turn subdivided into three stages, I, II, and III. Most of these stages have since been subdivided into early and late substages, A and B, and sometimes C. Thus, the first part of early Late Minoan time is designated "Late Minoan I A" or just "LM I A." The chronology is given in Table II. (It should be borne in mind that archeologists are not in complete agreement as to the precise dating of the various stages and substages and their correlations with Egyptian chronology.)

These Minoans were a seafaring people. At the height of their power they controlled the whole Aegean region politically, while their economic influence extended throughout the whole eastern Mediterranean and as far westward as Sicily. They were a peaceful people, depending entirely on the sea and on their fleet for defense; their palaces and towns were not fortified. They built no warships as such, but their merchant vessels carried warriors when action was necessary. The major Minoan communities were centered around a number of splendid palaces, each the seat of a priest-king. These local rulers were united in some form of confederacy, with the priest-king at Knossos supreme over all.

The standard of living enjoyed by the Minoans was probably higher than in some parts of Europe today. Even the smaller houses were two or three stories high, with wide windows, courts, and often several kitchens. Such refinements as toilets with arrangements for flushing, with drainage and sewer systems, were known in some of the palaces. The luxury goods made in the palace workshops by highly skilled artisans were traded as far away as Egypt and Syria. Ceramic art was very highly developed, the pottery reflecting gracious living. Fresco and vase painting and gem carving (particularly seal stones) reached a perfection not again achieved for almost a thousand years. Other artisans sculpted in ivory and stone, or wrought exquisitely in gold, silver, and bronze. Minoan fashions in dress, like their fashions in pottery and other design, changed throughout the years as fashions do, but what comes to mind as "typical" dress for Minoan ladies of high fashion is the tiered skirt falling from a tiny cinched-in waist, worn with a jacket which covered the arms to the elbow but bared the breasts completely. Coiffures were elaborate. The upper class men were clad mainly in a brief kilt with codpiece, but their bare chests were adorned with massive jeweled collars and their heads with fancy headdresses.

TABLE II

Minoan and Egyptian Chronology
(900-2000 B.C.)

B.C.	CRETE	EGYPT Dynasty
1000	Sub-Minoan	XXI
1100	Late Minoan III C	XX
1200		Raamses III
1300	Late Minoan III B	XIX
1400	Late Minoan III A	Ikhnaton
	Late Minoan II	XVIII
	Late Minoan I B	Tuthmosis III
1500	Late Minoan I A	
1600	Middle Minoan III B	
	Middle Minoan III A	HYKSOS
1700	Middle Minoan II B	XIII
1800	Middle Minoan II A	XII
1900	Middle Minoan I B	
2000		

(Second Palaces / First Palaces)

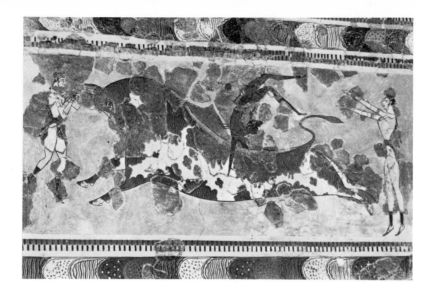

PLATE 38. *The "Toreador fresco" from the palace at Knossos, depicting the popular Minoan sport of bull-leaping. (Courtesy of the Archeological Museum, Heraklion.)*

The religion of the Minoans seems to have centered around a mother-goddess and other personifications of natural forces, and public worship made use of natural sanctuaries such as caves and groves. They built no elaborate temples, erected no heroic statues of gods or heroes; trees, poles, and pillars were worshipped as the visible abode of deities. Small cult rooms in the palaces and homes probably were used for royal or private worship. Music and dancing played a large part in their religious ceremonies, and perhaps in everyday life too. A popular spectator sport was bull-leaping (Plate 38), in which both young men and young women were sent into the ring to perform dangerous acrobatic feats with a bull, such as grasping the horns and flipping over the animal's back, vaulting or somersaulting over it, leaping on it, and so on. Boxing was very popular, also javelin throwing, and the Cretans were noted for their skill with the bow down to classical times.

Several times in its long history the palace at Knossos was damaged by earthquakes, and several times it was rebuilt, as were some of the other palaces. The destruction at the end of Middle Minoan III was particularly widespread and marks the close of the era of the "First Palaces." A very much more drastic change took place at the end of Late Minoan I. Suddenly, disaster seems to have struck everywhere at once. With one exception, every palace and every mansion, and in some cases whole towns, were reduced to ruins;

none of the great houses were ever rebuilt, and some of the towns were never reoccupied [174, 175]. Kato Zakros, Palaikastro, Mochlos, Pseira, Gournia, Nirou Khani, Mallia, and Amnisos on the coast lay in ruins—also Tylissos, Sklavocambos, Hagia Triada, Phaistos, and other sites inland, at elevations up to six hundred to seven hundred feet above sea level (Fig. 34). In many cases the

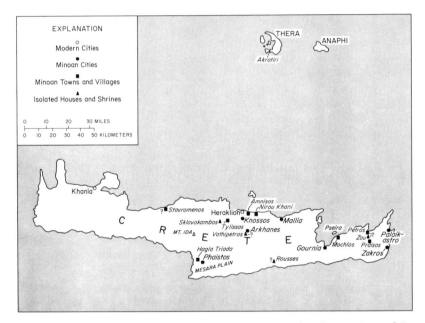

FIGURE 34. *Crete, showing sites destroyed at the end of Late Minoan I B. (After Hood, 1970 [115] and Luce, 1969 [143].)*

destruction appears to have been accomplished by, or at least completed by, fire. People returned to Tylissos, to Gournia, to Palaikastro, and to some of the other settlements, and made their homes again beside the ruins of the great houses. Other sites, such as Pseira and Mochlos, were abandoned forever. At about the same time Kydonia, a hitherto relatively unimportant town in the western part of Crete (approximately on the site of modern Khania), began to assume importance, and new settlements appeared in the western part of the island. Of all the palaces, only Knossos remained standing and continued to be occupied; but even there the style of life changed appreciably. Large rooms were divided into smaller

apartments. Designs in pottery became less tasteful, more flam-
boyant. And most important, the language now spoken at Knossos
was not Minoan.

One of the most significant archeological developments of the
1950s was the decipherment of the script called Linear B. Two systems of writing have been found on Crete. The older script, Linear
A, apparently represents the language of the Minoans and is found
throughout Crete. It seems to have been used mainly for practical
purposes in everyday life, such as inventories, rather than for literary purposes, and its development can be traced from early
pictographic beginnings to a syllabic form. The Linear B script
appeared at Knossos after Late Minoan I. It is known on Crete only
from Knossos, but later appears on the mainland. Even before either
was deciphered, it was recognized that Linear B was the same syllabic script as Linear A, but used for a different language—just as
the Latin alphabet can be used for French, German, English, and
other languages. The decipherment of Linear B in 1953 [32]
shocked some, delighted others, and surprised many [1:35-36]—
including its decipherer, Michael Ventris—when it turned out to
be an archaic form of Greek. This proved that Crete had been taken
over by Myceneans from the mainland. Those who had maintained
that the Myceneans were little more than barbarians and not very
powerful at the time were disconcerted; how could such a nation
have conquered the highly superior Minoans? Those who had insisted that the Myceneans were advanced enough and powerful
enough were delighted; in subduing Crete, the Myceneans demonstrated their supremacy. Like Linear A, Linear B was used only for
inventories and commercial transactions. It has been suggested that
when the Myceneans took over, they instructed the palace scribes
to adapt their script to Greek, but that "the results left something
to be desired and so the Linear B syllabary remained an inadequate
medium for the Achaean tongue. The new script then spread from
Knossos to the mainland, still no doubt in the hands of Minoan,
Knossian clerks and only for use in the palaces" [146:28].

Linear A has, not yet been deciphered, at least to the same
general satisfaction as Linear B. It definitely is not Greek. Is it an
Indo-European language, and if so, which one? Luvian and Hittite
[42, 189] have been suggested. Or is it Semitic [83, 84], as has also
been proposed? When Linear A yields its secret it should help to

throw light on the origin of the Minoans, but it is not their beginning but their end which concerns us here. For from the very beginning of our acquaintance with the Minoans, the suddenness and thoroughness of their collapse, plummeting all at once from the pinnacle of power to the status of a minor dependency of Mycenae, has baffled all those concerned with ancient history. None of the usual explanations for the decline and fall of a great nation seemed to fit here. The decline was far too abrupt to be laid to increasing decadence. Displacement by an invading horde could be ruled out, for no foreign culture supplanted the Minoan, it merely deteriorated. The change was so inexplicable that the suggestion was even seriously offered that perhaps the Minoans simply tired of their role as leaders of the Aegean world and handed over the reins to the Myceneans in a "passive renunciation of power"! [155:46]

After personally experiencing the 1926 earthquake, Evans concluded that the chief breaks in continuity of Minoan civilization at Knossos must have been due to "these cataclysmic forces of nature, here always latent" [58:320]. The obvious change in the style of living at Knossos after Late Minoan I B he attributed to its occupation by the common people, as a result of uprisings of depressed elements of the population who took advantage of the chaos following the earthquake. He further suggested that the feeling of insecurity induced by repeated destructive earthquakes, at intervals of a generation or two throughout a large part of Crete, might have induced people to emigrate, and may have encouraged overseas conquest and led to the wholesale recolonization of the mainland by men of Minoan stock [58:321]—for at the same time that the Myceneans established themselves as the rulers of Crete (or what was left of it), there appears to have been wholesale emigration to the mainland and corresponding depopulation of Crete. Aggression from Knossos (which so far as Evans knew in these pre-decipherment days, was still occupied by Minoans) could account for the destruction of any palaces spared by the earthquakes.

The Greek archeologist Spyridon Marinatos was not satisfied with this explanation. Only a disaster of greater magnitude than any earthquake could explain such wholesale destruction, and in particular, the failure to rebuild. On the basis of his excavations at Amnisos, where he found smooth, sea-borne pumice in the ruins of the palace, he turned to Santorin for the explanation. In a paper entitled "The Volcanic Destruction of Minoan Crete" [148], published in 1939, he proposed that it was the effects of the Santorin tsunami, together with an earthquake (which was still needed to account for the destruction at inland sites) which had dealt the death-blow to Minoan supremacy. The havoc wrought on the coast by the tsunami should have been sufficient to ruin the economy of a people dependent on the sea; their merchant fleet and harbor installations would have been destroyed, together with buildings of all sizes and grandeur, and thousands of people would have been drowned.

This idea was at first greeted with great skepticism. Granted, the total ruin of the coastal cities with great loss of life would certainly have crippled a nation dependent on maritime trade for its prosperity, and on ships for its defense. But as we have seen above, the damage caused on Crete by tsunami and by earthquake could not have been as severe as Marinatos assumed. Even a Krakatau-like or larger tsunami would not have wreaked equal havoc at all points on Cretan shores, nor would it have affected any Minoan ships that were at sea at the time, or in foreign ports which were spared any serious consequences. No volcanic earthquake could have been felt strongly enough on Crete to bring down whole cities or even individual edifices—although shock waves from the most violent explosions might have caused the same kind of damage as a minor earthquake here and there, and the possibility of a serious tectonic earthquake within months of a few years of the climax of the eruption cannot be ruled out. But a tectonic earthquake so severe that it toppled palaces and mansions and humbler dwellings throughout eastern Crete should not then have spared Knossos, and although fire sometimes follows an earthquake, a holocaust should not have ensued in such a large proportion of cases. And why did the Minoans not rebuild after this earthquake (if it was an earthquake), rather than leave the most habitable part of Crete in droves?

Two American oceanographers, Dragoslav Ninkovich and Bruce Heezen, apparently provided an answer to the last question specifically, and gave an additional reason for the suddenness of the decline of Minoan civilization, when they proved, in the study of deep-sea cores mentioned above, that a substantial part of Crete had been covered by ash from the Santorin eruption [170]. Crete is a dry land, and most of what rain does fall is concentrated in the autumn and winter months. The pattern of distribution of the Minoan tephra in the deep-sea cores (see Fig. 33) suggests that it fell in summer, when the prevailing winds are mostly from the northwest. The Icelanders in 1947 were able to salvage most of their farmland from the Hekla ash, with the aid of bulldozers and tractors and considerable help from wind and rain—for that summer the rainfall was double the usual amount [238:15]. The rains also quickly washed the fluorine-contaminated ash off the grass. But on Crete, even double the normally scant summer rainfall would not have been much help to crops smothering beneath an ash blanket. Nor, of course, would the Minoans have had the Icelanders' advantage of modern machinery to help clear their land—only the manual labor of a disheartened populace. Even if the ash contained no harmful substances such as fluorine or sulfur dioxide, its effects could have been serious wherever it accumulated in substantial thickness—and such accumulations would have been greatest in the low level places, those most likely to be under cultivation.

When the consequences of a damaging ashfall were taken into account, Marinatos' theory stood on more solid ground and gained much wider acceptance. Still, problems remained. Was the ashfall thick enough to necessitate abandonment of the land for many years, or only temporarily? The tentatively estimated ten centimeters (four inches) might have hurt field crops and pastures in low-lying areas, but should not have been enough to kill olive trees or vines, and would surely have been washed from hillside pastureland in one or two seasons of torrential winter rains. More important, exactly when did the eruption occur? To be the immediate cause of the decline of Minoan Crete, the eruption—or at least its most violent paroxysms, which showered ash on Crete, and the tsunami-generating collapse of the caldera—should have occurred at the time of the general destruction on Crete, at the end of Late Minoan I B. How can we determine exactly when it happened relative to the stages of Minoan culture?

In this case radiocarbon dating is not precise enough to be of much use. The best available radiocarbon date, obtained on a tree buried in the pumice on Thera which was growing at the time of the catastrophe, gives an age of 1456 B.C. (±43 years) [143:64].

Well and good; this is in excellent agreement with the archeological date of 1450 B.C. for the end of Late Minoan I B (see Table II). But the "±43 years" means that it could have been as early as 1493 B.C., which is in equally good agreement with the archeological date for the end of Late Minoan I A, or as late as 1407 B.C., by which time the Myceneans were already established at Knossos. Furthermore, if the date is calculated on the basis of the "preferred" half life of carbon-14 (see Appendix B), it turns out to be 1559 B.C. (±44 years), and if we correct for the effect of radiocarbon fluctuations in the atmosphere (see Appendix B), both dates come out even older—1673 and 1771 B.C., respectively! [159] However, the radio-carbon date of archeological objects also become older when the correction is applied, so the *relative* values are not affected. While it may be a shock to archeologists to be told that the Minoan stages may be a couple of hundred years older than they thought on the basis of correlations with Egyptian chronology, from a geological standpoint the ultimate validity of Marinatos' theory depends on whether or not there is a causal connection between the eruption and the demise of Minoan Crete, whatever the absolute dates turn out to be. To avoid additional confusion, therefore, we will use the archeological dates.

To try to pinpoint the time of the eruption more precisely than the radiocarbon method allows, we must turn to the archeological evidence, and here some conflict with the geologic evidence comes in. Buried under the thick tephra on Santorin itself are the ruins of a flourishing Minoan colony. Systematic excavations on Therasia in 1866-67, at Balos on Thera in 1870, and at two sites near Akrotiri on Thera in 1870 and 1899, respectively, revealed a prosperous community, with solid and stylish dwellings in which luxuries were by no means unknown. Only one skeleton, that of an old man, was found; he apparently was killed by debris falling in an earthquake. No further excavations were made until 1967, when Marinatos began a new dig at Akrotiri. So far this site has yielded a group of very impressive mansions, with interior stone staircases, breath-taking frescoes, and splendid pottery, mostly of local manufacture. The walls of several buildings at the new site seem to have collapsed before the pumice buried them, presumably in an earthquake. No more bodies have been found, and this, together with the absence of jewelry or other objects of value, indicates that the settlement was abandoned before the pumice fell.

The problem is that so far, no pottery later than early Late Minoan I B, represented by one single find in 1972, has been found anywhere on Santorin. This suggests that Thera was abandoned not long after the end of Late Minoan I A, whereas the widespread destruction of Crete definitely occurred at the end of Late Minoan I B. The Late Minoan I B period was the time of greatest artistic achievement in Cretan pottery, culminating in the "Marine Style," so called because subjects such as octopus, nautilus, dolphin, and starfish were used in its designs. These exceptionally fine vases are not numerous and were obviously produced by a small group of artists in a relatively short time, while other workshops still continued to turn out typical Late Minoan I A styles. At present the earliest date the archeologists are willing (reluctantly) to accept for the end of Late Minoan I B is 1470 B.C. (they prefer 1450 B.C.), and the latest for the end of Late Minoan I A, 1500 B.C., for at least a generation is needed to account for the amount of Marine Style pottery known.

The earlier interpretation of the tephra layers on Santorin (two violent phases separated by a long interval of mild intermittent activity) [194:122-127] fitted perfectly with this theory: the first phase caused the abandonment and burial of the Minoan colony on Santorin, while the second and more violent phase was responsible for the general destruction on Crete a generation or more later. But if, as is most likely in the light of more up-to-date geologic knowledge, there was only one brief violent phase, it is crucial to know whether it occurred at the *beginning* of Late Minoan I B, as the lack of mature Marine Style pottery in the ruins of Santorin would seem to suggest, or whether it was more or less simultaneous with the general destruction on Crete at the *end* of Late Minoan I B. In the first case, the eruption could not have been the immediate cause of the downfall of Minoan Crete; in the second, it is difficult to explain the near-absence of Marine Style on Santorin.

The idea that Marine Style pottery was well developed at the time of the eruption but had by-passed Thera almost completely can be ruled out as very unlikely, for it has been found at smaller and less sophisticated settlements, on Keos for instance, and as far away from Crete as Rhodes. Another suggestion, that the inhabitants of Thera were frightened away by an earthquake and stayed away for thirty years [174], is thoroughly unrealistic. For one thing, that would be completely out of character; even when entire communities have been flattened, people in the Mediterranean lands have always returned and rebuilt within a few years if not immediately; and judging from the history of Knossos and the other palaces up to the time of their destruction, the Minoans were no different in that respect from the people who live in the Mediterranean area today. Moreover, the earthquake could not have been an unusually severe one, judging from the extent of damage at Akrotiri. And although there appears to be some archeological evidence of very temporary reoccupation after the earthquake, presumably by those seeking valuables left behind, there also is geological evidence that the shock occurred immediately before the first major paroxysm, or simultaneously with it, for earthquake cracks filled with fresh pumice have been reported [115]; had they been open for any length of time before the pumice fell, they would also contain other debris washed or blown in before the eruption. And although purely volcanic earthquakes strong enough to cause damage are definitely rare, if one did occur in this unusually violent eruption, it is most likely that it would have been the shock accompanying the first major paroxysmal outburst.

Even after the discovery that the Santorin tephra was present in deep-sea cores all around eastern Crete (which to a geologist is proof enough that it must have covered that part of the island too), some archeologists were still skeptical. Why did they not find traces of the ash in their excavations? For the simple reason that they were looking for a visible layer of volcanic material, forgetting that habitations are the very places where volcanic ash is most likely to have been cleared away first, swept outdoors to be carried away by the agents of erosion. Any particles surviving to this day would be very few; lodged in crevices in walls or pavements, microscopic in size, mingled with local dust and soil, they would be completely indistinguishable to the naked eye. In the open, the steep rocky terrain of Crete is anything but favorable for the preservation of undisturbed remnants of the original ash cover, and the only lake on Crete in whose sediments a recognizable layer might have been preserved, Lake Kournas, lies outside the area of probable distribution of the Minoan tephra. Only systematic collection of samples, to be examined later under the petrographic microscope, could reveal the presence of tiny shards of volcanic glass which are all that one could reasonably expect to find today of the Minoan ash.

In 1971 my husband and I made such a search, and we found particles of the Minoan tephra in Cretan soils, all the way from the eastern tip of the island to as far west as Heraklion (the farthest west we looked), and also in samples collected from crevices in buildings occupied in Late Minoan I [261]. The presence of such particles in the soils confirms the evidence of the deep-sea cores, but of course tells us nothing specific about when the ash fell on Crete with respect to the Late Minoan stages. Likewise, their presence in buildings destroyed in Late Minoan I B—at Arkhanes, Mallia, Gournia, Zakros [261], and Pyrgos [28]—tells us only that it fell before the destruction; but if those buildings were occupied throughout Late Minoan I time, it could have fallen as early as Late Minoan I A just as easily as immediately before (or simultaneous with) the destruction. Only the presence or absence of ash particles in Late Minoan I A levels which were destroyed and buried under rubble before Late Minoan I B, and thus effectively sealed off from "contamination" in later events, could throw definite light on the timing of the eruption. And our samples from Late Minoan I A levels, all collected at Kato Zakros and at least one of them (freshly dug) sealed before Late Minoan I B, not only *do* contain particles of Minoan tephra, but contain them in greater abundance than any other levels sampled, including Late Minoan I B, from the same site.

If the archeological identification of the levels from which we collected these samples is correct, and if Minoan tephra particles show up in other samples of freshly dug, sealed Late Minoan I A from other sites, then the conclusion will be inescapable that the eruption began and ended around 1500 B.C., totally destroying the Minoan colony on Santorin but not putting an end to the Minoan civilization on Crete and other islands. Does this mean that we must totally reject the theory of the volcanic destruction of Minoan Crete and start all over again to find the reason for its bafflingly abrupt decline? As the immediate cause, yes. But what about the long-range effects of an occurrence of this magnitude, as the ultimate or at least a contributory cause of the decline?

The following is a hypothetical sequence of events as they *might* have happened. Since it can never be proved or disproved, it perhaps should be classed as science fiction, but like all good science fiction it is grounded on scientifically plausible possibilities—and in view of the fact that most authors dealing with this question have vastly overestimated the probable effects of the eruption, it just possibly could be closer to the mark than many another explanation that has yet been offered.

A series of mild earthquakes began to be felt by the people of old Stronghyli. At first they were not particularly alarming, for earthquakes had been part of normal existence there as long as anyone could remember. However, the shocks increased in frequency and severity far beyond the usual level of seismicity, and everyone wondered what they might portend. No other islands, including the mother island, were being subjected to any abnormal seismic activity. One or two families, relative newcomers to the colony, packed up their belongings and returned to their former homes. The vast majority offered prayers and sacrifices to their deities and went about their business as normal. Why leave? What part of the Minoan realm was *not* subject to earthquakes? True, Stronghyli was a little more unstable than other areas, but look at the recent shock which had shaken parts of Crete, doing some damage at Knossos and Amnisos.

Before long it became only too apparent that their island was indeed different from its neighbors. High on the flanks of the majestic peak which crowned Stronghyli, goatherds reported that there were places where the ground had become hot and was giving off steam and foul-smelling gases. Could it be that the volcano was not extinct, as they had always believed?* As the emissions of vapor and the frequency and severity of the tremors increased, so did their apprehensions. Family after family packed up their valuables and took passage on the first available ship. Among those valuables were their best pieces of pottery decorated in the new style which had just begun to be imported from Crete.

One night eerie flames suddenly flared over one of the spots where hot vapors issued from the ground, from a crack encrusted

---

* The deep soil underlying the Minoan tephra on Santorin is proof that the volcano was quiet for thousands of years before that eruption.

with an old yellowish deposit.* This was taken as a sign from the gods, and a general exodus began the next day. And none too soon. Before all who wished to quit the island had time to do so, the mountaintop burst with a muffled roar. A plume of steam and tephra rose rapidly into the air, and as it ascended it assumed a shape which some compared to a gigantic stone-pine tree, and others to a still more gigantic mushroom. More explosions followed, at intervals of a few minutes. Against the ash-laden eruptive cloud, chunks of rock (looking like black specks as seen from the settlements nestled on the lower slopes) could be seen as they were hurled high into the air, to fall back into the crater or roll down the slope. When darkness fell the scene was even more terrifying. The eruptive cloud was luridly illuminated at its base, reflecting the fires within the mountain, and the ejecta, which appeared dark by day, described fiery parabolas against the blackness of the sky and mountain. The smaller ones faded to black before they hit the ground, but the larger one landed still aglow and traced a fiery path as they rolled down the slope. Vegetation near the summit was stripped of its leaves, and the red-hot bombs often started fires in the shrubbery.

Now there was panic. No time to take anything but the most valuable and most portable belongings! Jewelry, of course, and metal pots, but only the most precious ceramics. Every available ship, large or small, was pressed into service, and as the refugees reached other ports the call went out for more ships to complete the evacuation. The last people to leave the island were nearly hysterical with fright, for by then the explosions were stronger and more frequent, sizable pumice bombs were falling among them, necessitating that they protect their heads when they left the shelter of their houses, and the rain of ash was so heavy that the sun was dimmed.†

* About seven weeks before an eruption of Ebeko volcano in the Kurile Islands, which began in February 1967, an increase in the temperature of gaseous emissions caused spontaneous combustion of sulfur around fumaroles [158]. While it is not essential to the "plot" being developed here, it just possibly could have happened at Santorin, and I include it because it provides a nice dramatic touch.

† Up to this point our hypothetical sequence of events is based on an analogy with the reaction of the people of Saint Pierre, Martinique, in 1902. There, the first signs of the impending eruption of Mont Pelée were noted on April 2, in the shape of new fumarolic activity high on the side of the peak. Toward the end of April ash was falling continuously but lightly, yet sightseers were still climbing the volcano to have a look into the crater. It was not until a few days before the final calamity of May 8 that the violence of the activity increased to frighten-

Akrotiri and the other communities on Stronghyli had become ghost towns—but as in the ghost towns of our own West, there were a handful of old-timers who stubbornly refused to leave. The worst is over, or almost, they claimed; and for a time it looked as though they might be right. The violence of the explosions began to diminish appreciably, and the intervals between them became longer and longer. After a couple of weeks of comparative calm, reported by passing ships, a few people returned to the island to take stock of the situation. They found that while the activity was now confined to the crater and appeared to pose no threat, or even discomfort, to anyone keeping a respectful distance, things were still happening inside the mountain. Loading themselves with as much as they could carry of their abandoned possessions, they sailed back to their new refuges and reported that it would be premature to return as yet. Also during this lull, bands of looters appeared on the scene, attracted to the island like vultures by the knowledge that there was much to be had for the taking in the deserted towns and villages. Before long nearly everything of value except the frescoes on the walls had disappeared; only the everyday pots were left, particularly those too large to cart away conveniently. The few who had remained on the island were helpless to stop the pillaging and found it prudent to hide when the marauders were present.

Before long the volcano cleared its throat again, and the explosions resumed. Now the tempo of activity quickly accelerated to a climax. In a rapid series of explosions of a magnitude so far unparalleled, whose roar was heard in Crete and on the mainland, Santorin spewed forth countless tons of ash and pumice, nearly all of which fell on the volcano and in the sea in the immediate vicinity. The first of these major explosions was accompanied by a shock strong enough to topple the walls of several buildings (killing one old man as he tried to rush outdoors), and opened cracks in the ground which quickly filled with pumice. All the settlements on

---

ing proportions and large numbers of people prepared to leave the city. Their misguided governor, however, in an effort to calm their fears (and, as has been suggested, to keep them there at least until after an election scheduled for May 10 [25:103-4]), took up residence in the city with his wife and at the same time posted guards on all roads leading out, to turn back those who were not sufficiently reassured by his own example [139:8]. He and his wife were among the thirty thousand or so who perished. The Stronghylians were more fortunate inasmuch as nobody prevented them from leaving if they wished to go.

Stronghyli were buried, some completely, some partially. Pumice lumps and fine ash found their way into buildings through every window and door opening. Roofs, weakened by the seismic shocks, caved in under the increasing load, often bringing walls and upper floors down with them. The handful of people who still remained on the island fled toward the beach, but few reached it; most of them lay where they fell in the choking darkness; the others found that the sea offered no sanctuary, for it was so littered with pumice that they coud not launch their boats. So perished the last inhabitants of Stronghyli.

On Crete, the effects of the first major outbursts were disconcerting and inconvenient, but not particularly harmful. The thundering roar of the distant explosions was frightening, and even more so were the shock waves which cracked walls and even brought down some old mud-brick buildings which were not in good repair. Most awesome was the sight of the huge dark cloud which rose rapidly on the horizon and dimmed the sun, even hiding it completely for a few hours at a time. A rain of very fine ash particles fell from this cloud, getting into eyes, hair, food, and clothing, sifting into all corners of dwellings, and dusting the crops in the fields and the vines and olive trees. To placate the gods, who all too obviously were seriously disturbed about something, the Cretans flocked to their sanctuaries and made the sacrifices and offerings deemed appropriate to the occasion. One of the offerings took the form of burying little cups containing pieces of pumice beneath the threshold of the room set apart for religious purposes.* Scattered lumps of pumice had been drifting ashore since the beginning of the very violent phase of the eruption, just a few days before, and its source was well known.

The grand climax of the eruption made all that had gone before seem insignificant by comparison. The display was terrifying in places so distant that none could guess the origin of the phenomena they were witnessing; on Crete and the islands nearest to Stronghyli, where many were aware that an eruption was in progress, the impact was no less terrifying, and physically uncomfortable or even

---

* Such votive deposits have been found in a large house at Nirou Khani, east of Amnisos. A tripod vase containing lumps of pumice recently found in a fourteenth century B.C. level at Kydonia is thought to reflect a late survival of a cult inspired by the eruption [115].

dangerous in addition. First there was a series of ear-shattering roars, louder than the muffled booms they had been hearing so far (the loudest were heard as far away as Scandinavia and well into Asia and Africa). Shock waves damaged poorly constructed buildings—even buildings made of stone—up to several hundred miles away, and toppled perfectly sound mud-brick upper stories all over Crete and other Aegean islands. Following hard upon the shock waves there descended a darkness which on Crete swiftly became so dense as to be almost palpable. A very extensive area was blanketed with tephra. Eastern Crete was covered with fine ash deep enough to smother low-growing crops and pasture grass, and to weigh down tree branches and vines. Branches broke under the weight, if they were not shaken free of their load as it accumulated; the vines suffered similarly. Even Lower Egypt was sprinkled lightly but noticeably with volcanic dust.

When the air cleared once more, it could be seen that the shape of Stronghyli had changed. The top of the tall cone appeared to have been beheaded as if with a sword. Steam still rose from fissures on its flanks, but not very vigorously. The entire island, shrouded in a deep mantle of grayish-white tephra, was a desert and would remain one for generations. The sea around it was littered with floating pumice, so thick that one could have walked upon it, had there been anyone to try. Navigation was impossible in the vicinity for some time, until banks of pumice gradually broke away and drifted from the scene to form weird floating islands of all sizes.

Eventually curiosity overcame trepidation, and a party of Mycenean sailors stopped to explore the grim landscape. Some of them who climbed the truncated cone found that the apparently flat top was really the rim of a large bowl-shaped depression. In its bottom steam rose from several vents aligned along fissures. Except for several small avalanches of rock and tephra, which slid down the steep sides of the bowl even as they watched, there was nothing to break the utter stillness except the sound of their own voices, which they instinctively hushed. Suddenly affrighted, they hastened back to their comrades. Their comrades, they discovered, were likewise subdued; they had been unable to find any trace whatever of the bustling port city they knew from previous voyages. Awed, they put back to sea, and on the homeward voyage mused about the ways of the gods—for what else could explain such complete and utter annihilation except heavenly displeasure?—and recalled how the Stronghylians had been the most arrogant and overbearing of the Minoans, and that was saying a lot. And as they drew farther and farther away from the accursed island, their spirits rose and they even began to gloat about the fate of a part of their rival nation.

Meanwhile, back on Crete there was consternation. The most productive lands, the valley floors and plains, were smothered under several inches of volcanic dust, and attempts to salvage at least part of the year's crops were thwarted by heavy out-of-season showers; for while those scoured the hillsides, they dumped more sediment on the plains. Sometimes, on steep slopes, the saturated ash slumped down in oozy mudflows which buried deeply whatever grew at the foot. Hillside olive groves, orchards, and vineyards did not suffer very much, however, particularly if their branches had been shaken free of the ash load as it accumulated. There was considerable belt-tightening that autumn and winter, for not only were grain and vegetables in short supply, but the livestock grew lean without adequate pasturage. But nobody actually starved, even if the poor people were reduced to using substantial amounts of carob in their diet.*

With the coming of spring, spirits rose. The new crops, planted where the ash cover had been washed thin by the rains or where the land had been laboriously cleared over the winter, promised to be as good or better than the previous year's. New grass sprouted through the rapidly thinning ash, particularly on the hillsides, and the cattle and sheep and goats began to lose their scrawny look. Windy days, to be sure, were annoying, for then the fine particles of sharp volcanic dust got in the eyes, and blew into the houses; but as sooner or later the dust found its way into the sea, on the whole the wind was regarded as an ally.

But while life on Crete gradually resumed its normal tempo, beneath the surface there lurked a strong feeling of uneasiness. The people of Stronghyli had been typical prosperous Minoans. If the gods had been so displeased with them as to force them into exile and obliterate their cities and villages, might not something equally

---

* During the occupation of Crete in World War II, when the Germans commandeered most of the local food supplies, the people subsisted by using large amounts of carob, or St. John's Bread, which grows more or less wild on Crete.

dire be in store for Crete? The uneasiness was heightened when visitors to Stronghyli during the ensuing months reported that all was not yet quiet on that island. The depression in the top of the peak was deepening, slowly but noticeably. More scrupulously than ever before the people observed the rites of their religion, and the priestesses even invented new rites involving pumice, which kept drifting ashore long after the eruption had ended. Never had the Minoans so wholeheartedly obeyed the strictures of their priest-kings and priestesses. Crime dropped to an all-time low. The proud walked more humbly. And the artisans in certain palace workshops outdid themselves in creating more beautiful objects, especially the vessels for use in religious ceremonies. Although it had been permanently disastrous for those who had lived on Stronghyli, and temporarily a hardship for those who had been showered with volcanic ash, the eruption might even be said to have ushered in the golden age of the Minoans. Should anyone be inclined to backslide, there was always the island of Stronghyli, slowly crumbling into its summit depression, to serve as a constant reminder. In time the central depression became lower than the sea surrounding the island, and one day, after a particularly large chunk had subsided, the caldera rim was breached and the sea rushed in, creating a bay where once the highest peak had risen. Within an hour a sea wave—just like the waves which often followed earthquakes—inflicted substantial but not irreparable damage at a number of points along the north coast of Crete. Such waves became abnormally frequent during the next few years, and had the Minoans realized it, each one followed hard upon the engulfment of another chunk of old Stronghyli.

And if the visible disturbing signs were not enough to keep alive the uneasiness underlying the outward prosperity of the Minoans, there were the Myceneans, long jealous of Minoan dominance of the Aegean (which among other things involved their paying an annual tribute to Knossos), who lost no opportunity to taunt the Minoan traders about the fate of the proud Stronghylians and suggest that some unnameable doom awaited Crete too. So the years passed. Many who could recall the days of noise and darkness and terror, and the ensuing winter of privation, were no longer among the living. Those who had been mere children at the time now had young children of their own. And then the gods struck again.

Beneath the Aegean, deeper than usual and more convulsively than usual, the earth shuddered. The shock was felt all over Crete, with varying intensity, but as usual even in the strongest shocks, damage was concentrated in one area—in this case, easternmost Crete. Zakros suffered the most. In the midst of a banquet in the palace, a sharp foreshock sent nobles and slaves alike rushing outdoors just in time. Panic-stricken, they watched the walls topple and flames from overturned oil lamps turn into a holocaust which swept through the ruins. The fire was so hot that limestone blocks were calcined to white lime, and chunks of mud brick were partially melted.

Enough was enough! Completely dispirited at this further indication that the deities had again singled out their country as the object of wrath, Minoans began to emigrate to other lands. A few followed up connections they had made in the course of their trading and went to countries as distant as Egypt and Asia Minor, but most of them turned to Mycenae, where skilled craftsmen in particular were always in demand. These Cretan refugees provided an infusion of new blood and new ideas which added impetus to the just-beginning emergence of Mycenae, the ancestor of our own western civilization, as a great power in the Aegean.

Those who remained on Crete found themselves plagued by increasingly daring bands of marauders, who took advantage of the growing confusion and depopulation to strike at those points on the coast where the fine houses still stood. Often the looters ended by putting the torch to the ruins. Word of the rapidly deteriorating situation spread, and finally a force from Mycenae landed at Amnisos and marched to Knossos, where they met only token resistance from the remnant of the priest-king's guard. Almost with relief the last Minoan monarch surrendered and was escorted to Mycenae, where he lived out the rest of his days in luxurious semicaptivity. The local rulers who still occupied palaces which had survived earthquake and marauders were younger and less weary of the struggle to reestablish order. They fought the invaders valiantly but vainly, and were either slain or enslaved, and their palaces burnt to the ground. The common people adjusted to their new rulers, their day-to-day existence not recognizably changed. The Minoan potters who had not left, striving to please the nouveau-riche tastes of their new masters (or possibly ordered to), developed the florid "Palace Style," which was reserved for the exclusive use of Knossos; elsewhere the potters continued to produce the styles which had been popular since Late Minoan I A. When Knossos was finally overthrown years later, Crete was nothing more than another Mycenean dependency, and not a very important one at that. The Minoans, once the leaders of their world, had vanished from the front pages of history, and Crete has never since emerged from the obscurity of the back pages.

9 /

Lost Atlantis Found?

Tʜᴇ ʟᴏɴɢ-ʙᴜʀɪᴇᴅ ᴍɪɴᴏᴀɴ ᴄɪᴠɪʟɪᴢᴀᴛɪᴏɴ ʜᴀᴅ scarcely been brought to light by the excavations at Knossos before the possibility of a connection between Crete and Atlantis was suggested—and promptly forgotten. The current revival of interest is essentially a corollary of the proposed volcanic destruction of Minoan Crete. But before we look into the merits for the case for an Aegean Atlantis, we must know the source of the Atlantis story and its essential features, and understand the reasons for rejecting an Atlantic site.

Contrary to the prevalent notion, Atlantis is not folklore at all— that is, it is not part of the oral traditions of any culture anywhere on earth. Every single mention of it stems from one and only one written source,* Plato's *Dialogs,* specifically the *Timaeus* and the

---

* There are a few earlier writings thought by some to be references to Atlantis. Homer and Hesiod both refer to the ancient idea that somewhere far to the west, beyond the boundaries of the inhabited earth, lay a paradise for departed heroes, the Garden of the Hesperides with its golden apples. (The Avalon of Arthurian legend seems to be an echo of the Hesperides; in fact the name *Avalon* is from the Celtic meaning *apple*). This western paradise of classical mythology, it has been argued, is a reference to the Atlantic islands such as the Azores, or even to the New World, glimpsed by storm-tossed mariners who somehow managed to find their way back to tell the tale. This may well be the origin of the notion of a paradise beyond the western sea, but the two essential elements of the Atlantis tale—a superior nation suffering a catastrophic end—are conspicuously absent. They are also missing in Ogygia, the island in the middle of the sea where Odysseus dallied with Atlas's daughter Calypso in the course of his wanderings, which also has been suggested as a pre-Plato reference to Atlantis.

Critias [182]. The *Timaeus* purports to record a conversation between Socrates, Timaeus (a scientist), Critias (a historian), and Hermocrates (a general), in which they discuss the nature of the universe. Critias tells the tale of Atlantis, which Solon, the great lawgiver of Athens who lived about two hundred years before Plato, is supposed to have told Critias' grandfather. In his youth Solon visited Saïs in the Nile Delta, then the capital of Lower Egypt (see Fig. 38, Chapter 10). There he discoursed with learned priests, and in the course of the discussions found that he knew very little of his own country's ancient history. To encourage the priests to tell tales of antiquity, Solon began to tell of the earliest event of which the Athenians were aware, the Deluge of Deukalion. Upon this one of the more ancient priests exclaimed: "O Solon, Solon, you Greeks are always children . . . your souls are juvenile, neither containing any ancient opinion derived from remote tradition, nor any discipline hoary from its existence in former periods of time." Was Solon not aware that the Deluge of Deukalion was only the last of a series of catastrophes, and that the Athenians were descended from a noble race who lived long before it? Of the many and mighty deeds of these old Athenians, whose city was founded nine thousand years before Solon's conversation with the priests, the most outstanding was the defeat of a warlike power from the Atlantic Sea.

This warlike power came from an island larger than Libya and Asia put together. The island "afforded easy passage to other neighboring islands; as it was likewise easy to pass from those islands to all the continent which borders on the Atlantic Sea." (For better understanding of this geography the reader is referred to Figure 35. To the Greeks of Solon's time the world consisted of "Europe" and "Asia"—the latter consisting of Asia Minor and North Africa— separated by the Mediterranean Sea, all surrounded by the "Ocean Stream," which in turn was surrounded by a continent. By Plato's time the ocean was known to be much wider than a stream, and had been named the Atlantic, but so far as is known, it had not yet been explored by the Greeks.) When the kings of Atlantis attempted to conquer and enslave all the lands of the Mediterranean area, the ancient Athenians led the fight against them and "procured the most ample liberty for all those of us who dwell within the Pillars of Hercules." Subsequently there were great earthquakes

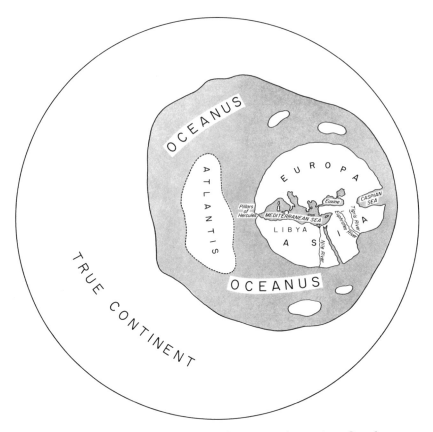

FIGURE 35. *The world as it was known to the ancient Greeks.*

and deluges, and in the space of a day and a night the ancient Athenian race was "merged under the earth," while Atlantis disappeared beneath the sea, leaving only unnavigable shoals to mark its site.

In the *Critias*, a dialog with the same four participants, Critias gives fuller particulars about the history, geography, and religion and culture of Atlantis, and then goes on to tell how the Atlanteans had gradually degenerated until Zeus thought their wickedness should no longer go unpunished; whereupon he called the gods together and said. . . . and there the dialog ends abruptly. Some think Plato died before finishing it, while others believe he started it earlier but shelved it in favor of other matters and never got around to it again.

Was Plato describing a place he believed to be real, or was he making it all up to prove a philosophical point? If he believed it to be real, how much of his description can be trusted? Works on Atlantis are of three main kinds: some try to prove Plato's description to be literally true; some allow for distortions in time or place or both; while others flatly refuse to accept it as anything other than fiction. If we ignore those who base their arguments on occult revelations and similar fantasies and do not let themselves be confused by mere facts, we find that all those in the first category, and many of those in the second, have sincerely quoted what they believe to be valid scientific evidence. All too often, however, they have either drawn grossly incorrect conclusions from established scientific facts, or have based their arguments on outdated scientific theories. Truly scientific analyses, on the other hand, have invariably supported those who consider Atlantis to be fiction, and who at best will admit that it could be based in part on real facts known to Plato [5, 44, 227].

The popular concept of Atlantis was crystallized in a book by Ignatius T. T. Donnelly, a self-educated man of extremely wide-ranging interests. *Atlantis: The Antediluvian World* [51], first published in 1882, was the first of his several very successful books. (Another which has had equally lasting impact, in a very different field, was *The Great Cryptogram*, in which he tried to prove that Shakespeare's works are liberally sprinkled with cryptographic clues indicating that Sir Francis Bacon was the real author. As with Atlantis, the idea was not original with Donnelly, but his book provided the springboard from which adherents of the view took off to become a persistent cult.) Donnelly's concept of Atlantis was that of a continent in the Atlantic Ocean, inhabited by a superior race, which existed until about 11,500 years ago and foundered in a great cataclysm, the survivors making their way to other lands and taking their superior culture with them. Donnelly began by stating thirteen propositions which he then proceeded to try to prove. His "proofs" were based mainly on comparisons of Old and New World civilizations, the worldwide distribution of flood traditions, and supposed references to Atlantis in Old World mythologies. Only three of these propositions are directly relevant to our geologically-oriented discussion:

1. That there once existed in the Atlantic Ocean, opposite the mouth of the Mediterranean Sea, a large island, which was the remnant of an Atlantic continent, and known to the ancient world as Atlantis.

2. That the description of this island given by Plato is not, as has long been supposed, fable, but veritable history.

12. That Atlantis perished in a terrible convulsion of nature, in which the whole island was submerged by the ocean, with nearly all its inhabitants. [51:1]

In connection with propositions 1 and 12, Donnelly invoked geologic arguments to show that (a) vast landmasses once existed where Plato located Atlantis, and (b) it was possible for a continent to be destroyed in a night and a day. The first argument was reasonable in the light of the geologic knowledge of that day. Deep-sea soundings had just revealed the existence of a ridge in the middle of the North Atlantic Ocean, which was given the name Dolphin Ridge (it is the northern part of the Mid-Atlantic Ridge), and similarities in the flora and fauna on opposite sides of the South Atlantic had already suggested that some sort of connection once existed between them. But Donnelly's inferences from the geologic evidence available to him concerning the second point were dubious in the extreme. Essentially his arguments boiled down to this: Continents have risen and sunk throughout geologic time; islands have been known to vanish overnight (Santorin is given as an example; Krakatau, of course, had not yet erupted and collapsed in 1882); ergo, there is nothing improbable in the statement that the continent of Atlantis was submerged as suddenly as Plato claimed.

Donnelly ignored the fact that while continents have indeed risen and sunk, they have done so with exceeding slowness. If a large island or continent were to be submerged, the process would take many millennia; it could hardly happen suddenly enough to leave a memory of great disaster, which is what Donnelly (like many still today) believed the Deluge tradition to be. Floods, to be sure, might inundate extensive areas rather quickly, but even the most widespread flood imaginable would eventually drain away and leave the land still above water. The only geologic forces capable of producing sudden and permanent submergence of land are earthquakes and caldera eruptions, but they involve only a few dozen square miles at a time, at most. All in all, there is simply no geologic mechanism which can submerge a large land area quickly and for all time.

The late Lewis Spence, a Scottish mythologist, attempted to reconcile Plato's account with the geologic facts as he understood them, by changing the time of Atlantis. He recognized the glaring inconsistency between the definitely Bronze Age culture described, and its destruction 11,500 years ago; for the Bronze Age did not begin until about 5,000 years ago. In 1925 [222:42-43] he suggested that the destruction of Atlantis was only the final event in the breakup of a huge continent which formerly occupied all or much of the North Atlantic. That continent, he claimed, began to disintegrate in late Miocene time, "owing to successive volcanic and other causes." Two large remnants, Antillia and Atlantis, persisted until 25,000 years ago, connected by a chain of islands. The West Indies represent the remains of Antillia; Atlantis continued to disintegrate until disaster finally overtook its last remnant in about 10,000 B.C. The piecemeal disappearance of Atlantis gave rise to successive waves of migration over a long period of time, one of the migratory pulses coinciding with Plato's date for the destruction of Atlantis. The superior race was the Cro-Magnon, whose advanced Stone Age culture resembles that of ancient Central America.

By Spence's time geologists were more or less agreed that there had once been a connection between the Old and New Worlds. The similarities in the pre-Mesozoic fossil flora and fauna on both sides of the South Atlantic were too startling to be coincidental. Some postulated narrow land bridges of more or less temporary existence, like the present isthmus of Panama, while others still considered the "Dolphin Ridge" to be a foundered large landmass. In 1912 a striking piece of "evidence" turned up which was hailed as proof that the "Dolphin Ridge" had once been above sea level. An eminent French geologist [234] gave a lecture in which he stated that pieces of rock brought to the surface in 1898 by a grappling iron seeking a broken telegraph cable were of basaltic glass (tachylite), similar to rock formed in the 1902 eruption of Mount Pelée. It had been observed that where the Pelée lava had solidified in the open air it was glassy in texture, but where it had solidified under a cover of other lava it formed typical fine-grained basalt. This he interpreted as proof that the glassy rock from the ocean floor had formed under atmospheric pressure, and therefore an area now lying about two miles deep must have once been above sea level.

Unfortunately for this argument, it is not the pressure but the rate of cooling which governs the crystallization of magma, a fact which was already known at the time the lecture was given. If magma is chilled very quickly, crystals have no time to grow and the resulting rock will be glassy. It makes no difference whether the chilling is due to contact with the air or with cold water deep under the sea. If it cools more slowly, like the Pelée lava which was not exposed to the air, it will be finely crystalline, and if it cools very, very slowly deep in the earth it will give coarse-grained rocks like gabbro, in which the individual crystals are visible to the naked eye. Glassy crusts have been observed on basalts recovered from depths to 17,000 feet or more off the Hawaiian Islands [163], so there is no reason why the tachylite dredged up from the "Dolphin Ridge" could not have formed precisely where it was found.

Land bridges over which plants and animals could cross the ocean are still postulated by some geologists today, though their numbers grow steadily fewer. The presence of land bridges recent enough to have permitted the migration of human beings is easily demonstrable only where the regions are still nearly contiguous, such as the opposite sides of the Bering Strait or the eastern Canary Islands* and Africa. Land bridges across the Atlantic, however, are much more conjectural and there has never been any general agreement as to where they might have existed. In any case, even if they did once exist, the connection was severed by the end of Mesozoic time 70 million years ago, for after that the floras and faunas on both sides of the Atlantic evolved independently. Thus land bridges, even if they once were present, are no help at all in trying to prove the existence of land in the middle of the Atlantic as recently as 11,500 years ago.

The more we learn about the Atlantic Ocean floor—and our

---

* The Canary Islands have long been a contender for the honor of being the site of Atlantis, and Spence thought that the final wave of migration from Atlantis came via those islands. But the Canaries as a whole do not represent the foundered remnants of a larger landmass; on the contrary, the western islands of the group have been built up from the sea floor by volcanism during the last tens of millions of years [212].

knowledge has increased by leaps and bounds in the last several decades—the more difficult, even impossible, it becomes to find any place where land connections could ever have existed. We know that the earth's crust underlying the oceans is essentially different from that underlying the continents; it is much thinner, and the "granitic" layer present under the continents is missing.* There is no large area of continental-type crust now submerged beneath any ocean, and there are no submerged narrow strips of that type of crust linking the opposite sides of any ocean.

Moreover, current geologic concepts do away completely with the need for an intervening land of any kind at any time. Today the theory of continental drift, which until about a decade ago was not taken very seriously by many geologists in this country, is completely respectable and is winning adherents daily. According to this theory (first formulated in 1924 by an Austrian geologist, Alfred Wegener, and independently and almost simultaneously developed by Frank B. Taylor, an American) the continents were once all joined together in one or perhaps two supercontinents, which broke up into separate blocks which have been drifting apart slowly for the last 70 million years. Continental drift easily explains why the coastlines on opposite sides of the Atlantic fit together when cut out and moved about on the globe; it explains why systems of rocks in Africa and South America match in structure as well as in age and fossils; it explains why some differences in the paleomagnetic† directions of rocks older than Tertiary can be eliminated by shifting the continents back toward one another.

---

* Most of what we know about the parts of the earth which are too deep to investigate at first hand by drilling (which means all but the very outermost "skin" of the earth) is inferred from the behavior of seismic waves as they propagate from earthquakes to seismic stations all over the world, or from large or small explosions to seismic instruments on land or on ships. Seismic waves travel with different speeds in different materials, and the velocities for all kinds of rock have been measured in the laboratory; thus the speed of seismic waves propagating in different layers of the earth is an indication of the kind of rock constituting those layers.

† While certain rocks are being formed, magnetic particles in them are magnetized in the direction of the earth's magnetic field prevailing at the time, and under normal conditions will retain that magnetization even if the position of the rock in relation to that field is subsequently changed. There are various forms of this *remanent* or fossil magnetization. The most useful, because it is very stable, is *thermoremanent* magnetization, the kind acquired by lava as it cools. Volcanic

Continental drift was not accepted with enthusiasm at first, largely because the mechanism proposed was not plausible. It pictured rigid blocks of lighter material called "sial" (essentially granitic in composition, high in aluminous silicates) floating like icebergs in denser, somewhat plastic "sima" (essentially basaltic, high in iron and magnesian silicates). Today, even though the mechanism is still not yet fully understood, two independent lines of evidence make it difficult to believe that the continents have *not* moved relative to one another. The latest refinement of the theory is the concept of "plate tectonics," which has revolutionized geologic thought. According to this view, the lithosphere (the earth's crust and part of the mantle—see Fig. 36) is divided into a number of plates (whose exact thickness is a matter of opinion, but which may be from sixty to two hundred miles thick), created at the mid-ocean ridges and destroyed in the oceanic trenches, which slide about on a weak layer in the upper mantle, which, judging from the fact that seismic waves are retarded in it, is in a more plastic state than the material above and below it—presumably because it is partially melted. This weak layer is called the *asthenosphere*.*

Where plates collide, things happen geologically. If two plates carrying continents come together, the result is a folded mountain range; the Himalayas, for instance, formed when India drifted northward against Asia. Where an oceanic plate bumps against a continent, as the northwest Pacific does against Asia and the southeast Pacific against South America, it plunges under the lighter continental plate; its path of descent is marked by an inclined plane of

---

rocks like basalt, which have a high proportion of iron-bearing minerals including magnetite, the most easily magnetized of all minerals, will acquire a readily measurable remanent magnetization. Sedimentary rocks containing a lot of iron oxide, such as red sandstones, are also useful because their individual iron oxide mineral grains tend to align themselves with the geomagnetic field as they are deposited. If samples of such rocks are carefully collected, noting their exact position in space, the direction of their remanent magnetization can then be measured in the laboratory after suitable preparation, and from that direction, the position of the magnetic poles of the earth at the time of their formation can be plotted. Paleomagnetic measurements also record polar wander, which is a different kind of movement of the magnetic poles; in polar wander there is movement of the whole earth (or at least its outer "shell") with respect to the pole—or of the pole with respect to the earth. In the case of polar wander alone, there would be no relative movement of the continents with respect to one another.

* The prefix *astheno-* is from the Greek meaning *weak*.

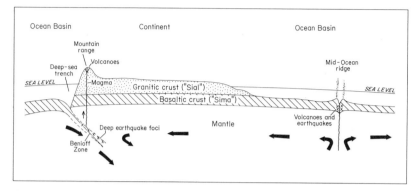

FIGURE 36.· *Continental drift according to the new global tectonics. New material from the earth's mantle rises along a mid-ocean ridge, causing the plates (consisting of the crust and uppermost mantle) to move apart. In this case, a westward-moving plate, carrying a continent (which could be South America), collides with an eastward-moving oceanic plate (Pacific Ocean plate), and the latter dives downward beneath the continent. Physical and chemical reactions between the down-plunging crustal material and the mantle generate volcanism and earthquakes. (After Takeuchi and others, 1967 [232].)*

earthquake hypocenters called (in honor of the American seismologist Hugo Benioff) a *Benioff zone*, by an oceanic trench offshore, and by mountains crumpled at the edge of the continent. When the down-plunging material descends deep enough it melts and is very slowly recycled back to a mid-ocean ridge in a sort of conveyor-belt type of action thought to be driven by convection currents. The association of active volcanism with descending plates is not accidental, but is due to degassing of the downgoing lighter material. Not all plates collide head on, and some rotate a bit as they move around. The northeast Pacific is side-swiping the west coast of North America and diving under the Aleutian Islands, tearing Baja California away from Mexico and producing the well known San Andreas fault and its branches, with which much of California's earthquake activity is associated.

The two lines of evidence which converge to support the idea that the ocean floors really are spreading [48] are "magnetic anomalies" and potassium-argon dating of rocks from the ocean floor. Although the *direction* of the earth's magnetic field—that is, the geographic location of the magnetic poles, which do not quite coincide with the geographic poles—has not changed appreciably since the beginning of Tertiary time, its *polarity* has reversed itself —that is, the north and south magnetic poles have become interchanged—many times, rather abruptly, all throughout geologic time. The last known reversal occurred between 13,500 and 17,500 years ago [37]. How and why it happens is not known, though it has to be bound up in some way with the processes in the earth's fluid outer core which create the main geomagnetic field, as the earth spins on its axis like a gigantic dynamo. Whatever the reason, piles of lavas erupted in various geologic epochs show alternations of normal and reversed polarity* which can be correlated all over the world.

It was predicted [259] that if the ocean floors are spreading apart from the mid-ocean ridges, as the plate tectonics concept postulates, then the alternating paleomagnetic directions of the rocks created in successive epochs of alternating magnetic polarity should show up in the pattern of magnetic field anomalies as stripes paralleling the mid-ocean ridges on either side; and this has been confirmed by magnetic surveys over the oceans (made with airborne or ship-towed magnetometers). The rate of sea-floor spreading has been estimated as at least one centimeter per year, some estimates running as high as eight, and the rate apparently has not been constant. The evidence of the magnetic anomalies has been supported by the results of radioactive dating of rocks collected from the ocean floor at different distances from the Mid-Atlantic Ridge; the ages obtained were those predicted on the basis of the polarity evidence, showing a uniform increase away from the center on both sides of the ridge.

By this time, no further arguments should be needed to prove that no large landmass or island could ever have existed in the Atlantic Ocean, certainly not in man's time on earth. But there is one more line of evidence which drives yet another nail in the coffin of an Atlantic Atlantis, from the relatively new field of paleoclimatology [20]. By means of biogeochemical techniques based

---

* "Normal" magnetization means that the north and south magnetic poles lie near the north and south geographic poles, respectively, as they do today; "reversed" means that the magnetic north pole was near the geographic south pole and vice versa.

on the ratios of the stable (i.e., nonradioactive) oxygen isotopes in the shells of marine organisms, it is possible to determine the temperature of the water in which those organisms lived and secreted their shells. Most oxygen atoms have an atomic weight of 16, but a few—in atmospheric oxygen, about two out of a thousand—have an atomic weight of 18. From studies on living species it is known that the cooler the water, the greater the proportion of "heavy" oxygen (O^{18}) entering into the carbonate forming the shell. Thus, by measuring the relative proportions of the two oxygen isotopes in fossil shells we can calculate the temperature of sea water in past ages, which in turn tells us what the climate was like in those days. (This thumbnail sketch of the method is vastly oversimplified, but it serves to give a general idea of the principle involved.) At any rate, the paleoclimatic evidence shows that 11,500 years ago things were cool in the Atlantic Ocean in middle latitudes—where Atlantis was supposed to be—which is to be expected inasmuch as that was about the end of the last Pleistocene glaciation. The climate implied in Plato's description of Atlantis, however, is a mild climate much like that enjoyed by the Mediterranean area for the past few thousand years [57].

With the physical possibility of a former landmass anywhere in the Atlantic definitely ruled out on geologic grounds, all arguments for an Atlantic Atlantis which are based on cultural and linguistic similarities between the Old and the New Worlds collapse under their own weight. If such similarities are not just coincidental or purely imaginary, they must have some other explanation. (Those questions, however, are beyond the scope of this geologically-oriented discussion.) Either Atlantis existed elsewhere than in the Atlantic, or it did not exist at all outside of Plato's imagination.

By judicious selection of those parts of Plato's account which fit, and rejection of those which do not as distortions or exaggerations, a case can be made for almost any part of the world as the site of a historical Atlantis, and indeed it is hard to find any part of the world which has *not* been proposed at one time or another: the Arctic; various parts of Europe and the Mediterranean; North and South Africa; North, Central, and South America; Ceylon; and even the South Pacific have all been nominated. In many instances the choice of site reveals a good measure of chauvinism, the proponent trying to prove either that his own country was the real site of Atlantis, or barring that, that his own nation is descended from the Atlanteans and therefore superior to others. Spence, for instance, says:

> If a patriotic Scotsman may be pardoned the boast, I may say that I devoutly believe that Scotland's admitted superiority in the mental and spiritual spheres springs almost entirely from the preponderant degree of Crô-Magnon blood which certainly runs in the veins of her people. . . . England also, undoubtedly draws much of her sanity, her physical prowess and marked superiority in the things of the mind from the same source, and if much of her blood be Iberian, is not that too Atlantean? To an admixture of Crô-Magnon and Iberian blood we owe the genius of Shakespeare and Burns, Massinger and Ben Jonson. Milton, Scott, and, to come to our own times, Mr. H. G. Wells and Mr. Galsworthy are almost purely Crô-Magnon. . . .
> [222:230]

The Basques also claim descent from the Atlanteans and believe that their language, related to no known language today, is a survival from the original tongue of Atlantis. And the great seventeenth century Swedish scholar Olaf Rudbeck believed, as did many of his contemporaries, that his name would be remembered principally for a huge unfinished treatise in which he "proved" that Atlantis was the Scandinavian peninsula and Sweden the original home of civilization. In fact, however, he is remembered as the discoverer of the lymphatic system of the human body [112].

Any discussion of all the suggestions as to the alternative sites for Atlantis would alone fill a good-sized book. The reader who is interested in further details is referred to L. Sprague DeCamp's comprehensive work entitled *Lost Continents,* first published in 1954 and reissued in slightly revised form in 1970 [44]. His geologic discussions are thoughtful and fundamentally sound.*

---

* Most interesting is DeCamp's dismissal in 1954 of continental drift with these words: "Altogether, perhaps we had better put the Wegener theory on the shelf marked 'very doubtful' and leave it there for the time being" [p. 162]. In his 1970 edition the theory has been taken down from the shelf, dusted off, and displayed in all its new respectability [44:162].

Inasmuch as geologic thought has changed even more significantly since 1925, when Lewis Spence's theory was originally published, it is an interesting example of the reluctance with which cherished theories are given up to find that, in a posthumous reissue of Spence's work in 1968, his Atlantic site theory is merely

The same selection and rejection process which permits consideration of the numerous alternative locations for Atlantis must also be invoked in connection with an Aegean site. Why then should it be taken any more seriously than others? Because, for the first time, both of the absolutely essential elements of the Atlantis story—a superior civilization and a natural catastrophe—are involved.

The first person to think of the Minoans as the possible prototype of the Atlanteans was the classical scholar K. T. Frost, who, first in an anonymous letter to the London *Times* on January 19, 1909 and later in a more detailed exposition in which he acknowledged authorship of that letter [64], emphasized that the legend makes good sense when regarded as historical from the Egyptian point of view. To the Egyptians, the disappearance of the Minoans from the scene just when they seemed strongest and safest would have looked as though they had sunk into the sea. But skeptics who demanded a literally submerged Atlantis were not slow to point out that Crete is still very much above water. In 1928 a Russian named L. S. Berg [13] tried to locate Atlantis in the Aegean near Crete. The Aegean was formed in geologically recent times (Quaternary) by subsidence of a block of land ("Aegeis") which once joined the Balkan peninsula and Turkey; only its high points remain above water today, forming the Aegean islands. Berg thought that the memory of that former land could have been handed down to the Minoans, who in turn could have mentioned it to the Egyptians during their commercial contacts, and that the Egyptians could have derived from that a tradition of catastrophe. This theory has little to recommend it, for it requires that the end of Aegeis, which was very gradual and unremarkable, be remembered for tens or hundreds of thousands of years.

When Marinatos proposed his theory of the volcanic destruction of Minoan Crete in 1939, he was quick to recognize its implications for Atlantis. In 1950 he published a paper in which he argued that the Atlantis myth had probably grown from the fusion of various disparate episodes which occurred over a span of about nine hundred years, but had as its core "the destruction of Thera

---

restated, the reader being referred to the original work for the geologic arguments in its support—and those fifty-year-old arguments are called the ideas of "modern geology"! [221:53]

accompanied by terrible natural phenomena, felt as far away as Egypt," giving rise to "a myth of an island, beyond all measure powerful and rich, being submerged." As his paper was originally published in Greek (the English version was not published until 1969, for the First International Scientific Congress on the Volcano of Thera [149]), it did not receive very widespread recognition. The current interest must be credited largely to A. G. Galanopoulos, who, in a series of works beginning in 1960 [67-69, 72-76, 78], has endeavored to demonstrate that the Bronze Age eruption of Santorin can account not only for Atlantis but also for several other myths and semihistorical traditions—including the Deluge of Deukalion.

Marinatos' and Galanopoulos' views represent two quite different paths leading from Minoan Crete to Atlantis. Galanopoulos' route is based on a belief in the existence of a garbled historical record of the Minoans and the Santorin eruption, brought by Solon to Greece from Egypt in translation and handed down to Plato to be recorded by him some two hundred years later. To travel this route, then, necessitates that we find a plausible explanation for the differences between Plato's description and what is known concerning the Minoans and the Bronze Age eruption.

Let us start by following Galanopoulos' route, though not necessarily exactly in his footsteps; can we get from Crete and Santorin to Atlantis without meeting some insurmountable obstacle? The only way to find out is to consider in turn each relevant point in Plato's description (assuming for the moment that it *is* based on a real document) and judge whether it fits Minoan Crete and/or Santorin, or if not, whether some plausible explanation can be offered to reconcile the discrepancies.

1. According to the *Timaeus* ancient Athens was founded 9,000 years before Solon's time, and fought a war with Atlantis at some unspecified time afterward; according to the *Critias*, the war was fought 9,000 years before Solon's time. (If Plato was actually reporting, this inconsistency reveals a certain amount of carelessness on his part.)

One of Galanopoulos' fundamental arguments is that all figures over 100 (in the Greek text) have been exaggerated tenfold as a result of a translation error [72, 73, 76, 78], introduced when the tale was communicated to Solon by the Egyptian priest; one of them mistranslated the Egyptian word or symbol for 100 as 1,000. As will be seen when we get into other measurements, this idea has great merit inasmuch as it reduces *all* figures—whether they refer to time, place, or numbers of men and ships—to values which are consistent with what we know of Minoan Crete. Only someone who has had frequent occasion to compare translations with their originals can appreciate the beautiful simplicity and logic of the translation error argument. As a constant user, sometime editor, and now professional perpetrator of technical translations, I can assure the reader that an error such as Galanopoulos postulates is trifling compared to some I have seen, both published and unpublished.

Solon is known to have visited Egypt as a young man, though the exact time of that visit is not certain; it is thought to have been some time between 593 and 583 B.C., but may have been after 570 B.C. In any case, if we substitute 900 for 9,000 years, both the founding of Athens and the war with Atlantis fall within the limits of error of the available radiocarbon dates for the Minoan eruption, and one set of dates (after 1470 B.C.) is close to the archeological dating of the end of Minoan Crete.

2. "This power came forth out of the Atlantic Ocean . . . and there was an island situated in front of the straits . . . called the Pillars of Hercules; the island was larger than Libya and Asia put together. . . ."

Galanopoulos believes that Plato moved Atlantis to the Atlantic Ocean, realizing that an island of the exaggerated dimensions given could not possibly have been situated in the Mediterranean. He then offers arguments to prove that the "Pillars of Hercules" originally referred not to the Strait of Gibraltar, but to capes Malea and Taenarum (Matapan) (see Fig. 28) [76; 78:97]. It seems to me to be quite unnecessary to juggle the geography of the "Pillars of Hercules." Whether it was Plato himself or some ancient Egyptian who located Atlantis in the Atlantic Ocean, that meant putting it beyond what we know as the Straits of Gibraltar, and Plato naturally would have used the name by which those Straits were known to him.

3. "The island . . . was the way to other islands, and from these you might pass to the whole of the opposite continent. . . ."

There can be no doubt that Crete was a stepping-stone between North Africa, Asia Minor, the islands of the eastern Mediterranean, and Europe. As J. V. Luce puts it, "From the Egyptian point of view this is an accurate description of Crete as the gateway to the Cyclades and mainland Greece" [143:181].

4. A "great and wonderful empire" ruled Atlantis and several other islands and parts of the continent, and had subjected the parts of Libya within the Pillars of Hercules "as far as Egypt, and of Europe as far as Tyrrhenia."

Politically, the Minoans controlled Crete and many Aegean islands and part of the Greek mainland, while economically their influence extended at least as far as Libya, Egypt, and Sicily.

5. After the ancient Athenians had defeated the aggressive Atlanteans, ". . . there occurred violent earthquakes and floods and in a single day and night . . . all your warlike men in a body sank into the earth, and the island of Atlantis in like manner disappeared in the depths of the sea."

As shown earlier, heavy showers could have been caused by the Minoan eruption even at great distances from the volcano, due to the cloud-seeding effect of the ash in the high atmosphere, and in any case a tsunami generated by the collapse of Stronghyli (as we have been calling pre-eruption Santorin) certainly would have qualified as a deluge whenever or wherever its effects were felt. Concussions from the major explosions could have been mistaken for earthquakes, as in the case of Krakatau, or a tectonic earthquake could have occurred within a reasonably short time after the eruption. The climax of the eruption, including the caldera collapse, could have been accomplished in a day and a night. News of the sudden disappearance of a large part of a small island must have circulated around the Mediterranean for a time, probably losing nothing in the telling. On hearing of it, might not the Egyptians have put two and two together to make five, and linked it with the more or less abrupt disappearance of the Minoan traders from the scene? But whether or not they heard of the disappearing island, to what agency would the stay-at-home Egyptians have been more likely to attribute the erroneously inferred disappearance of all Crete than an earthquake? Even these days, as we have seen, people believe all kinds of impossible things about earthquakes.

6. ". . . the sea in those parts is impassable and impenetrable, because there is a shoal of mud in the way; and this was caused by the subsidence of the island" (*Timaeus*); "And afterwards when sunk by an earthquake, [Atlantis] became an impassable barrier of mud to those voyagers from hence who attempt to cross the ocean which lies beyond" (*Critias*).

There are no shoals in the Mediterranean which could be taken as the basis for this idea, but none are necessary. At the time of the Minoan eruption, as Galanopoulos has already pointed out [78: 135], the sea around Santorin must have been covered by a very thick blanket of pumice. Probably it was much thicker than that around Krakatau in 1883, and that was as much as ten feet thick. To the small ships of the Bronze Age the pumice-littered sea must have indeed been unnavigable, until such time as the floating pumice was gradually dispersed by wind and wave or, waterlogged, sank to the bottom. *If* the collapse of the caldera occurred while the pumice still blanketed the area thickly—that is, at the culmination of the eruption—it is even easier to see how the combination of sunken island and muddy shoals could have come into the description.

7. Atlantis was divided into ten portions, ruled by the descendants of five pairs of male twins born to Poseidon and a mortal woman, Cleito. The eldest, named Atlas, was "king over the rest." His descendants "had such an amount of wealth as was never before possessed by kings and potentates, . . . and they were furnished with everything which they needed, both in the city and country. For because of the greatness of their empire many things were brought to them from foreign countries. . . ."

This is completely consistent with what we know of the wealthy, sea-trading Minoans, ruled by a confederation of priest-kings subject to a supreme ruler at Knossos.

8. That the geographic details of Atlantis are not entirely clear in Plato's description is demonstrated by the fact that different translators differ as to the precise rendering of several passages. The part of the country where Cleito dwelt, where the Metropolis of Atlantis later arose, is described as "the fairest of plains and very fertile"; in the center of the plain was a low hill, apparently fifty stadia from the edge of the plain. But the location of the plain itself

is somewhat ambiguous: two translators agree that it was *towards* the sea, yet in the *middle* of the island; a third says it *bordered* on the sea and extended *through the center* of the island.

Around the hill, Poseidon had created perfectly circular alternating zones of sea and land. All translations agree that the central island, on which was built the royal palace and other edifices, was five stadia in diameter, the inner zone of sea one stadium, the next land and sea zones two stadia each, and the outer zones of sea and land three stadia each. Later kings of Atlantis bridged over the zones of sea and dug a canal from the sea to the inner harbor. The canal was three hundred feet wide, one hundred feet deep (three plethra and one plethrum, in the Greek), and fifty stadia long. If we draw a plan according to these specifications we get something like Figure 37.

The description of the surrounding countryside is less clear.

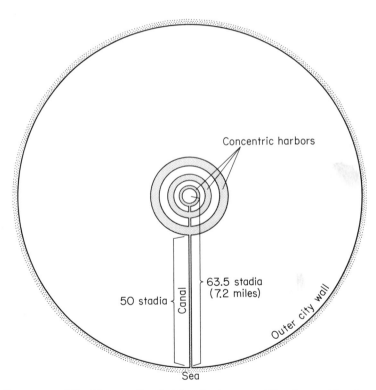

FIGURE 37. *The Metropolis of Atlantis, drawn to Plato's specifications.*

The Jowett translation [182] says that the whole country was very lofty and precipitous on the side of the sea, but immediately surrounding the city was an oblong level plain surrounded by mountains which descended toward the sea. According to the Taylor [182] translation every place near the sea was very elevated and abrupt, but around the city was a circular plain "itself circularly enclosed by mountains" which extended to the sea. And finally, the Loeb [182] translation tells us that the whole region "rose sheer out of the sea to a great height, but the part about the city was all a smooth plain . . . encircled by mountains which stretched as far as to the sea." All agree that the plain was three thousand stadia long and two thousand stadia wide. "This part of the island looked towards the south, and was sheltered from the north" (Jowett); or was it that "the whole island, likewise, was situated towards the south, but from its extremities was exposed to the north" (Taylor); or perhaps "this region, all along the island, faced towards the south and was sheltered from the northern blasts" (Loeb)?

No wonder that Galanopoulos argues that Plato was describing two different places! He suggests that the Metropolis of Atlantis was on the island of Stronghyli and the Royal City and its surrounding plain on mainland Crete, the Metropolis being the religious center and the Royal City the seat of government. In this view, the passages describing the Metropolis would apply to pre-collapse Santorin. The dimensions are appropriate; none of the measurements here involve figures of more than a hundred units (in the Greek), and thus they would not have been subject to the postulated translation error. *If* by any chance Santorin was already a caldera before the Minoan eruption [103], as Krakatau was before 1883, and *if* that caldera had been partially filled in by subsequent eruptions, just as the present caldera is being filled, only more so, then it is possible that Stronghyli did have a central depression with a low hill in the middle of it. But to fit the description of the Metropolis literally, the bottom of a conjectured pre-Minoan caldera would have had to be low enough that the sea could flow in through the artificially excavated channel (which would have been a tremendous engineering feat even in easily dug pyroclastic deposits), and that is not consistent with present interpretations of the pre-collapse configuration of the volcano. So far as we know, it rose from the sea to a central peak of unknown height. In any case, even if it ever existed there, to seek for traces of the Metropolis of Atlantis in the present bay of Santorin, as has been suggested, is unduly optimistic. Any traces of human works that might have remained on the sea floor after the collapse must long since have been buried in the lava and other eruptive products which form the present Kamenis (see Fig. 31).

On the other hand, the resemblance of the plain around the Royal City to the Mesara plain on Crete is undeniable. The Mesara plain is oblong and level, it lies on the south side of Crete and is sheltered by mountains on the north, and the dimensions are right when divided by ten. However, it is not Knossos which lies in the Mesara area, but Phaistos, and if Crete was the main island of Atlantis, surely Knossos was the Royal City. Luce dismisses the possibility that Plato's description of the Metropolis of Atlantis refers to Stronghyli, which he regards as just another small Minoan dependency, and locates the Metropolis at Knossos, situated on the low mound of Kephala in a rich and fertile plain [143:181].

9. Poseidon is said to have furnished the central island of the Metropolis with hot and cold running water in the form of two springs.

The hot spring has been taken as evidence in favor of Santorin as the site of the Metropolis, inasmuch as thermal activity is commonly associated with volcanic areas. As a general statement it is true that hot springs are associated with volcanic areas—but the association is with areas where there has been volcanism in the geologically recent past, like Yellowstone National Park; rarely (if ever) are they found *immediately* on or around still-active volcanoes. The thermal areas of Iceland, for instance, are related to geologically young but now essentially extinct volcanism, while hot springs are conspicuously absent on Hekla and the other live volcanoes. As for Crete, it is not and never has been volcanic. To the best of my knowledge, there are no hot springs there, not even of the kind related to high heat flow from the earth's interior (like those of Carlsbad, for example) rather than to dying volcanism.

10. Poseidon made "every variety of food to spring up abundantly from the soil." There was an abundance of wood for carpenters' work, and sufficient maintenance for tame and wild animals. All sorts of fragrant things—roots, herbage, woods, or

essences distilled from fruit and flower, also cultivated fruits, and chestnuts—"all these that sacred island brought forth fair and wondrous in infinite abundance."

This picture of a lush green land sounds little like relatively barren modern Crete, and there are no grounds for supposing that the climate was substantially different thirty-five hundred or so years ago. To be sure, pre-collapse Santorin might have been better watered if its peak was high enough to cause the clouds to drop their moisture on its flanks; and trees must have been more abundant on Crete, for its forests provided cypress for the Venetian navy as late as medieval times. And Crete is still noted for its aromatic plants and herbs: wild thyme grows everywhere, releasing its fragrance to the air when trodden underfoot; lichens common on Crete have been found in Egyptian tombs and may have been imported there for use in perfumes [143:141-142], and aromatic oils from Crete were used by the Egyptians in embalming [170]. Fruits are also cultivated in modern Crete. Grapes are grown everywhere, for wine, for the table, and for raisins; oranges are grown, but on irrigated land; fig trees and olive trees are ubiquitous. But large areas of Crete must always have been unproductive, even if the amount of land under cultivation in Minoan times was the same as at present. In 1948, 8 percent of the land was under cultivated crops and 10 percent in vineyards and orchards; 5 percent was lying fallow, 7 percent was grazing land, 2 percent was forest, and 48 percent was *madara*, bare land used for grazing (mainly goats and sheep); the remaining 20 percent was unproductive rugged mountain terrain. At least half the *madara* would have been virgin forest in Minoan times, but the wild crags and torrents would have been the same as today [118:41]. Then too, much of today's cultivation is made possible only by extensive terracing, and we have no way of knowing whether the Minoans made similar use of the steep hillsides.

11. In Atlantis "there was provision for all sorts of animals, both for those which live in lakes and marshes and rivers, and also for those which live in mountains and on plains"—and among the wild animals was "the largest and most voracious of all," the elephant.

Lakes and marshes are almost nonexistent on Crete, and there are only two permanent streams. And while it is no strain on the imagination to think that a few elephants could have been imported into Crete as curiosities, the mind boggles at the thought of herds roaming wild. An Egyptian fresco shows a Cretan envoy carrying an elephant tusk; it has been suggested [143:184] that the Egyptians could have assumed that the ivory came from Crete, whereas in fact the Minoans either hunted it in North Africa or traded it with Syria, which is known to have had elephants in the time of Tuthmosis III of Egypt, who reigned until 1439 B.C.

12. The Atlanteans "dug out of the earth whatever was to be found there," including *orichalcum*, "more precious in those days than anything except gold."

The mineral resources of Crete and the other Aegean islands are limited. The Minoans must have imported a large part of the metals they needed. Orichalcum, or "mountain bronze," is thought to have been an alloy of copper and zinc—in other words, brass—made empirically by adding powdered "cadmia" (zinc oxide) to molten copper; zinc as a metal was not known to the ancients [203: 155-158]. It is not known whether the Minoans knew orichalcum, or whether it is the same as the orichalcum mentioned in the description of Atlantis; for the latter is said to have "flashed with a red light," or "shone with fiery splendor," or "sparkled like fire," depending on which translation one reads—all of which suggests a reddish tinge which the known orichalcum did not have, for it was more whitish than ordinary brass.

13. The royal palace built on the site of Cleito's hill was a sumptuous one, each successive king striving to outdo his predecessor in its ornamentation, "until they made the building a marvel to behold for size and for beauty."

Knossos to a "T"?

14. Each of the land areas in the center of the Metropolis of Atlantis was surrounded by a wall made of stone quarried from beneath those areas. One kind of stone "was white, another black, and a third red." The wall around the center island or acropolis was coated with orichalcum, that around the next zone of land with tin, and that around the outermost zone with brass. Fifty stadia from the outer harbor "you came to a wall which began at the sea and went

all around" the Metropolis, "the ends meeting at the mouth of the channel which led to the sea."

White, red, and black stone are found on Santorin—but they are also found elsewhere in the Aegean. In any case, the Minoans certainly did *not* surround their settlements with walls, metal-clad or otherwise.

15. The acropolis which had been Cleito's hill was graced not only by the royal palace, but also by splendid temples, one to Cleito and Poseidon and another to Poseidon alone. In the latter there were gold statues, a huge one of the god himself standing in a chariot, and other images "dedicated by private persons," while around the temple on the outside were gold statues of "all who had been numbered among the ten kings" and their wives.

The Minoans definitely did not build temples or erect heroic statues, nor did they worship a sea god so far as we know; their religion seems to have centered around a mother-goddess, of whom they modeled or carved only small figurines up to one or at most two feet high.

16. Around the hot and cold springs provided by Poseidon, the Atlanteans constructed buildings and planted trees; "also they made cisterns, some open to the heaven, others roofed over, to be used in winter as warm baths; there were the kings' baths, and the baths of private persons . . . ; and there were baths for women. . . ."

These details are very consistent with Minoan palace life. The palaces were equipped with baths, not only for the use of the inhabitants, but also for ritual bathing of visitors before they entered the royal presence. At Knossos at least, the queen had her own elegant bathroom.

17. The plain around the Royal City was surrounded by a ditch one hundred feet deep, three hundred feet wide, and thirty thousand feet long, into which the streams coming down from the mountains were channeled to the sea. A network of intersecting canals fed by the main ditch criss-crossed the plain, by way of which wood was brought from the mountains and the produce of the plain was conveyed to the city. Twice a year they "gathered the fruits of the earth—in winter having the benefit of the rains of

heaven, and in summer the water which the land supplied, when they introduced streams from the canals."

Reduced tenfold, the dimensions of this system of canals fit the Mesara plain of Crete; but though the concentration of precipitation in winter is certainly consistent with a Mediterranean climate, the total rainfall implied in this description is much higher than that of Crete. Nor has any trace of an irrigation system been found in the Mesara plain or elsewhere, at least not yet.

18. For military purposes the Royal City was divided into sixty thousand lots ten stadia square, each with a leader who was required to furnish men and materials in case of war: one chariot for every six lots, plus horses and riders, heavy- and light-armed soldiers (the latter including archers, slingers, stone-shooters, and javelin-throwers) and sailors to make up a complement of twelve hundred ships. The other nine governments raised their quotas in slightly different ways which are not specified.

Reduction of the large numbers by a factor of ten brings the number of ships and chariots down to more credible proportions. Surely the priest-kings of Minoan Crete must have shared responsibility for ensuring the national defense, perhaps very roughly in the manner specified.

19. The kings of Atlantis met every few years to consult about their common interests and to pass judgment, and on that occasion they pledged their mutual allegiance in a ceremony which is described in some detail. Bulls were turned loose in the temple and hunted with only clubs and snares; the one caught was led up to a sacred pillar on which were inscribed the injunctions of Poseidon, and slain in the usual manner.

Whether or not the Minoans associated the bull with a sea god as did the Greeks—and so far as we know, they did not—they obviously regarded it as sacred, for the bull is one of the most conspicuous motifs in their culture. The netting of bulls is depicted on the Minoan gold cups found at Vapheio in Sparta. Representations of bulls trussed and ready for sacrifice on the altar are also known. Bull-leaping, although more of a spectator sport, may have had religious significance at least originally. The idea of the sacred pillar lends still another Minoan parallel to this passage, one of the most suggestive of all that the Atlanteans may have been the Minoans.

Assuming that after the Minoan traders had abruptly ceased coming to Egypt, some Egyptian priest compiled what information was available to him concerning Minoan geography, government, customs, and the circumstances of their demise as a great power, we would expect his chronicle to contain some information that was reasonably accurate, much that was distorted to varying degrees, and possibly even a few facts that did not pertain to Crete or the Minoans at all. Reviewing the arguments presented in 1 to 19 above, we find that 3, 4, 7, 13, 16, and particularly 19 are fully consistent with our picture of Minoan Crete; and that 1, 2, 5, 6, 8, 9, 17, and 18 can be made to fit, sometimes fairly easily, but sometimes only by tortuous arguments; however, 11, 12, 14, and 15 are quite contrary to everything we know. And underlying everything is the nagging question: If Solon was so impressed with the story of Atlantis that he intended to write an epic poem about it, why did he mention it to no one but his closest friend? Was it because he feared some other writer would steal the idea? But that still does not explain why Critias in his turn kept the story to himself from the age of ten, when he first heard it, until he was a mature man supposedly participating in the *Dilaogs;* for he too claims to have been very deeply impressed.

The ground traversed on the alternate route from Atlantis to Minoan Crete is underlain by the premise that Plato's description is essentially his own invention, but that he did not make it up out of whole cloth. In the time-honored fashion of authors everywhere, he wove into its fabric scraps of myth and tradition already at hand, and embroidered the whole with details based on his own experience and imagination. It we follow this path we need not concern ourselves with explaining the discrepancies between Atlantis and Crete, although it is interesting to speculate as to the possible sources of this detail or that; our main task now is to explain the more striking resemblances. How could Plato have known so much about the Minoans, who in his time were already the forgotten subjects of a mythical king?

The Theseus myth incorporates one of the two essential ingredients of Atlantis and some details of Minoan life besides. The Minoan domination of the Aegean is implicit in the fact that the mythical king of Crete exacted tribute from Athens. The youths and maidens who as a matter of fact danced dangerously before bulls (frequently, no doubt, with fatal finale) became in the myth that tribute, to be sacrificed to a bull-monster [201]. The Minotaur's labyrinth is quite obviously the palace at Knossos. Theseus' victory over the Minotaur reflects the fact that the Greeks gained the mastery over Crete. Finally, the fact that the Minoans were a sea power was plainly stated in Thucydides. Thus the myth and Thucydides, both well known to Plato, provide the picture of a sea-faring nation, stronger than any of its neighbors, in whose culture the bull played a prominent part, and which was taken over by the early Greeks. Then, if Plato deliberately chose the supposedly mythical people of Crete as the prototype of the Minoans, what would be more natural than that he should provide his imaginary country with a geographic setting resembling Crete, even if it was he who chose to move its location and change its dimensions? The geographic resemblances offer no obstacle whatever to our progress along this path.

In assigning a religion to his fictional maritime Atlanteans, it would have been equally natural for Plato to make them honor the sea god above all others, to model their worship on that of the Greeks—including the bull as a sacred symbol (already conspicuous in the Theseus myth), and the erection of temples and heroic statues—and even to trace their ancestry back to Poseidon himself. As for the details of the Atlantean bull ritual, however, it seems far too much of a coincidence for Plato's unaided imagination to have come so close to the mark. I wonder, is it possible that he could have seen representations of the Minoan ceremony on works of art like the Vapheio cups, objects lost to us but known to him as antiquities of Cretan, or even of unknown, origin?

But now what about the other essential ingredient of the Atlantis tale, the catastrophic disappearance? None of the myths and traditions concerning Minoan Crete that were available to Plato, so far as we know, contain even the glimmer of such an idea. Could some echoes of the Santorin event have survived, at least orally, to his time, not necessarily in connection with Crete at all?

The Bronze Age eruption and the collapse of the island of Stronghyli *must* have made some impression in its day, and it is indeed very surprising that the memory of that event apparently had completely faded by Plato's time. After all, the Klamath Indians seem to have preserved the tradition of a similar eruption which occurred more than 6,500 years ago (see Chapter 6). But there are several reasons why the memory of the more recent Aegean event would have dimmed faster. The people of the Mediterranean area were organized into far more complex societies than the Indians, and their way of life changed rapidly and often violently. In the interval between the Santorin eruption and Plato's writing of the *Dialogs* there had been repeated wars and other upheavals; attention must frequently have been diverted to "real and present dangers," and waves of migration brought new peoples into old settings. The Klamath Indians, on the other hand, followed their Stone Age pattern of existence in the same place for several thousand years, with virtually no change until the coming of the white man.

Probably if those nearest to the scene had known writing, some record would have been made of the Bronze Age eruption—perhaps only a cryptic one, like that of Old Mataram. But the Greeks did not yet have writing, and the Minoan Linear A script apparently was adequate only for limited household and commercial purposes. Moreover, the former inhabitants of Stronghyli were not engulfed. Perhaps if they had *not* had time to flee, the event would have been deemed more newsworthy by their contemporaries, particularly by those who lost friends and relatives in the disaster; in time there might even have emerged a myth involving divine retribution.

The memory of the eruption may have faded by Plato's time, but it did not wholly die. At least some echoes have reverberated down to our own time, but they have taken the form of myths and vague semi-historical traditions which have no obvious relationship to each other or to the eruption. This is entirely to be expected, inasmuch as the consequences of the eruption must have been experienced differently at different distances from the source of the disturbance, by people living in widely separated areas who had no way of comparing notes and thereby learning or deducing that the phenomena they witnessed had a common origin. And of all the effects, those of a Krakatau-like tsunami would have been more widespread, more calamitous, and therefore more memorable than any others. There is Deukalion's deluge which, as we have already seen, might be a memory of a Santorin tsunami; and Plato had Deukalion in mind when he wrote the *Timaeus*, for he has the Egyptian priest mention his deluge as one of a number of such catastrophes.

There are many other local Greek traditions of inundation, some of which also might actually stem from the Minoan eruption; Plato could have borrowed the idea of submergence from any or all of these. Apollodorus tells how Athena and Poseidon once contested for possession of Attica (a tale which Luce believes may reflect Minoan-Mycenean tensions [143:146]). Athena produced the olive tree, and Poseidon a spring. When Athena's was judged the more useful invention, Poseidon was so chagrined that he flooded all Attica. Pausanias tells that Poseidon lost a similar contest with Hera for possession of Argos, and thereupon flooded the Argive plain. Poseidon also contested with Athena over Troezen, the home of Theseus, with similar consequences; that wave was the "bull from the sea" which overwhelmed Theseus' son Hippolytos, as described in a novel by Mary Renault [200]. All three places are situated on the eastern Peloponnese, where a Santorin tsunami would have been strongly felt; however, the Santorin eruption cannot account both for the death of Theseus' son and for the take-over of Crete by the Myceneans implied in Theseus' overcoming the Minotaur as a very young man; either the oral tradition is inconsistent (which is more often the case than not) or the "bull from the sea" was an ordinary earthquake-generated tsunami at some later date.

Several other traditions cited by Luce are unmistakable references to tsunami, but not necessarily to the Santorin tsunami. In an area as seismic as that in question, there must have been many an earthquake-generated tsunami felt more severely in some places than a Santorin tsunami, even if the latter was the worst ever, on the whole. When Bellerophon, the youth who owned the winged horse Pegasus, rejected the advances of the wife of King Proteus of Argos, she falsely accused him of forcing his attentions on her. Proteus could not honorably kill one who had been a guest under his roof, so he sent Bellerophon to Lycia with a letter to King Iobates asking that the bearer be slain. But Iobates had accepted

Bellerophon as a guest before reading the letter and was likewise bound by the code of hospitality. He sent Bellerophon off to kill the fire-breathing and supposedly invincible chimera (see Chapter 4), expecting that the youth would be killed in the attempt. After slaying the monster, by shooting it from a safe height on Pegasus' back, Bellerophon prayed to Poseidon to punish Iobates, and the whole plain of Lydia was flooded by a huge wave.

Strabo reported that during the reign of Tantalus a great earthquake devastated Lydia and Ionia as far as Troy; marshes turned into lakes and a sea wave flooded the whole region around Troy. The unusual detail of the flooding of the marshes might be, as Luce suggests, the remote memory of the torrential rains which followed the eruption; but it is equally possible that the flooding of the marshes could have been caused by an earthquake which disrupted drainage à la Reelfoot Lake (see Chapter 5) and also produced a tsunami.

Excessive rain is specifically mentioned in a Rhodes tradition, in addition to a great "flood tide"—an extremely apt description of a tsunami—which wiped out the city of Cyrbe. After the destruction the land was divided between Lindos, Ialysos, and Cameirus, each of whom founded a city bearing his name. According to Luce there is archeological evidence at Triandra on Rhodes that after that Minoan settlement was destroyed a Mycenean colony was planted nearby, probably at Ialysos; the Minoan colonists rebuilt Triandra and coexisted with the Myceneans for a time, but eventually were overcome by the Myceneans at about the time of the final destruction of Knossos.

Yet another Greek flood tradition comes from Samothrace, telling of a deluge which ensued when the barriers dividing the Black Sea from the Mediterranean suddenly burst, and the Bosphorus and Dardanelles were carved out by the outpouring waters. In reality the passage from the Black Sea into the Mediterranean was cut at the end of Pleistocene time by normal erosion. Frazer calls this a "myth of observation," crediting it to the guess of some early philosopher who "rightly divined the origin of the straits without being able to picture to himself the extreme slowness of the process by which nature had excavated them" [63:170]. However, the historian Diodorus Siculus reported that in his time (he was a contemporary of Julius Caeser) the Samothracians were still sacrificing on altars set up around the island to mark the floodline of a great inundation from the sea [143:147], which points rather clearly to a real flood as the source of the tradition. Perhaps it was a Santorin tsunami, as Luce suggests; but again, it could equally well have been some other.

Luce [143:120] believes there is a specific reference to the Santorin eruption in Pindar's paean to Delos, composed at the request of the people of Keos. Pindar (522-428 B.C.) puts the words into the mouth of Euxantius, who extols the security of a small island over the intrigues and strife of a larger kingdom. Euxantius, a son of Minos, refused a seventh part of Crete in order to remain on his native Keos. As an omen against his leaving tiny Keos he refers to an earlier disaster: "I tremble at the heavy-sounding war between Zeus and Poseidon. Once with thunderbolt and trident they sent a land and a whole fighting force down to Tartarus, leaving my mother and all the well-fenced house." The heavy-sounding war could easily refer to the eruption. Krakatau's booms in 1883 were taken for cannon fire at several distant points, and in the days when spear and arrow were man's most formidable weapons, such sounds could only have been attributable to the conflict of gods. The land sent to Tartarus, Luce believes, could refer to the engulfed portion of Santorin, and the "fighting force" to the breaking of Minoan naval power as a result of the catastrophe.

Marinatos [149] believes that the sending of a host to Tartarus refers to the devastation of the coasts of Crete and elsewhere by the Santorin tsunami. I think it even more likely that it refers to the harm done on Keos itself. The apparently wanton destruction revealed in excavations at the Minoan settlement of Hagia Eirene on that island has been attributed to the Santorin tsunami [143:119]; if the land and host sent to Tartarus were the low-lying parts of Keos and their inhabitants, and if the "well-fenced" (or "strong-walled") house alone was situated above the level of destruction, then the juxtaposition of those ideas in the same sentence makes better sense to me. At the same time it is tempting to think that the host referred to was a military body and that Pindar was referring to the same event which provided the basis for Plato's statement in the *Timaeus* concerning "all your warlike men in a body" who sank into the earth at the same time as the submergence of Atlantis.

Then there is a subsidiary element in the Talos myth (see Chapter 6) which might be an echo of the Minoan eruption. Talos is said to have had a son named Leukos ("White"), who drove out the rightful king of Crete, murdered the king's daughter Kleisthera ("Key of Thera"), to whom he was betrothed, and destroyed ten Cretan cities. Leukos, suggests Luce, specifically represents the whitish Minoan ash which "covered the cities and fields of Crete after the 'death' of Talos himself" [143:150].

An episode in the chronicles of the Argonauts may also embody a memory of the Minoan eruption [143:150]. As they sailed on from Crete after overthrowing Talos, the Argonauts were enveloped by "black chaos coming down from the sky, or some other darkness rising from the inmost recesses of the earth," and lost their bearings completely. Jason prayed to Apollo, and the sun god guided them by the glint of his golden bow to the island of Anaphi (about 20 kilometers to the east of Santorin). I believe that some Greek ship, venturing close to Santorin during a lull in the Minoan eruption, was caught in a dense cloud of ash after an unexpectedly violent explosion*—not one of the climactic paroxysms, for at that stage the sea around Santorin would have been too littered with pumice to be navigable at all, but a minor one capable of causing a total blackout only locally and temporarily. Such a ship might land on Anaphi just by luck after losing its bearings, and having survived to tell of the adventure, the crew would provide a detail to be worked into the story of Jason and the Argonauts by later story tellers.

One final incident in the Argonaut saga indicates that someone was aware that the topography of Santorin underwent a drastic change shortly after the episode of the blackout. In an early stage of the homeward journey, the *Argo* had crossed Lake Tritonis (thought to be the Chott Djerid in Tunisia), where the local god Triton had presented Euphemus, one of the *Argo*'s pilots, with a clod of earth. As they were leaving Anaphi on the last leg of the journey, Euphemus tossed the clod overboard and it became the island of Kalliste, which we now call Thera [143:152].

There is a graphic passage in Hesiod's *Theogony* which de-

---

* In volcanic eruptions it is often observed that the strength of individual paroxysms is directly proportional to the time elapsed since the last one; for, other things being equal, the longer the interval, the greater the gas pressure that can be built up.

scribes the battle between Zeus and Typhon mentioned in Chapter 6: "And the heat from them both gripped the purple sea, the heat of thunder and lightning and of fire from such a monster, the heat of fiery storm-winds and flaming thunderbolt. And the whole earth and firmament and sea boiled. And long waves spreading out in circles went seething over the headlands, and unquenchable earthquakes broke out." Luce comments that "this passage could be interpreted as a classic description of a volcanic eruption complete with electrical storms, earthquakes, and tidal waves. But it must be admitted that there is nothing to locate it on Thera" [143:170]. I disagree—there is something which links it to the Minoan eruption rather than to any other, and that is the "tidal waves." Tsunamis are not normally associated with eruptions, usually only with submarine eruptions and with very few of those, at that. The one eruption of antiquity which may have generated a large tsunami, if the collapse of the caldera was sudden, is precisely the Bronze Age eruption of Santorin!

All in all, Greek mythology and traditions relating to the Minoans and to the Bronze Age eruption could have supplied the two basic ingredients of the Atlantis tale, but as two independent ideas, requiring fusion by Plato into the story he told. For although we know that the Minoans must have suffered more than any other people from the consequences of the eruption, no traditions specifically link the Minoans with catastrophe—at least, none surviving to our time.

If the two basic ideas of Atlantis, the powerful nation and a catastrophic submergence, actually came to Plato already juxtaposed, then we are almost forced to look back to Egypt as the source—but not necessarily to a specific document brought to Greece by Solon. If any written notice was taken of the Minoans and their virtual disappearance it would have been the Egyptians who took it. At the same time it is too much to expect that their records would be factually correct; distortions are more to be expected than otherwise. We cannot rule out the possibility that, directly or indirectly but somehow via Egypt, Plato could have learned the story of the catastrophic end of a great power. Because of the language barrier and other sources of confusion, he may not have recognized Crete, but he could have chosen the (to him) mythical Minoans as the prototype of his Atlanteans, either because the record or rumor purported to be about a Cretan people, or because (not surprisingly) it reminded him of the Crete he knew from the Theseus myth and from his own observation.

Everything considered, whether we believe in the reality of the Egyptian document "quoted" by Plato in the *Timaeus* and *Critias* (which I am inclined to doubt), thus following the route taken by Galanopoulos, or whether we believe that Plato manufactured Atlantis out of myths and traditions, some possibly coming to him via Egypt, which he altered arbitrarily and to which he added details from his own experience or imagination (which I am inclined to believe), thus following more or less along the path trodden by Marinatos, a case can be made for the derivation of Atlantis from Minoan Crete. The case is no more provable than any others yet offered, but is much more plausible than many another to date, and at least it is less *disprovable* than most. No argument other than tangible, independent documentary proof will ever settle the question, and it is not likely that such proof will be discovered even if the case is perfectly valid. Even if the Minoan civilization did *not* receive its death blow from Santorin, the case for a Minoan Atlantis is not destroyed; for the Minoans were a fact, and the eruption and collapse of Santorin were a fact, and it could have been Plato himself who put them together. In any event, we can expect that Atlantis will continue to provoke discussion and speculation for many years to come. I can think of no better way to sum up my own feelings on the subject than to quote a remark by Bruce Heezen, one of the authors of the paper on Santorin tephra [170] which put the theory of volcanic destruction of Minoan Crete on a firmer scientific foundation: "As for Atlantis, it's a lot of fun, and we just might be right."

10 /

Santorin, Egypt, and the

Eastern Mediterranean

WHILE CRETE AND OTHER AEGEAN ISLANDS received the full force of the various phenomena of the Bronze Age eruption of Santorin, some of its effects must have caused consternation or chaos in more distant parts of the Mediterranean world. Egypt, roughly five hundred miles from Santorin and right in the path of the prevailing northwesterly winds (see Fig. 33), could hardly have escaped experiencing a series of awe-inspiring manifestations. By analogy with Krakatau we can safely assume that at least lower Egypt was blacked out for a time by the ash cloud accompanying the major paroxysms, and that all Egypt should have heard the noise or felt the shock waves accompanying those tremendous explosions; moreover, any tsunamis could hardly have by-passed the Egyptian coast. Furthermore, Egypt was literate, so it would be very surprising if Egyptian writings contained no mention of unusual phenomena which could be interpreted as direct or indirect effects of the Minoan eruption. It would be even more surprising if the Egyptians who recorded these phenomena realized their connection with such a far-away event, even if they heard about it later.

However, there are not many records at all concerning that particular time in Egyptian history, which fell in the Eighteenth Dynasty (see Table II). It has been suggested [170] that little of the literature of the time has been preserved because Ikhnaton

(Amenhotep IV), the king who tried unsuccessfully to impose a monotheistic religion on Egypt, later ordered all earlier scriptures destroyed in an effort to erase all mention of the names of the ancient gods. There are, however, some later inscriptions and papyri which refer to events which may have taken place in the Eighteenth Dynasty.

Passages like these are encountered in the Hermitage papyrus in Leningrad:

> The sun is veiled and shines not in the sight of man. None can live when the sun is veiled by clouds.
> None knoweth that midday is here . . . his shadow is not discerned. Not dazzled is the sight when he [the sun] is beheld . . . he is in the sky like the moon.
> The river is dry, even the river of Egypt.
> The south wind shall blow against the north wind.
> The Earth is fallen into misery.
> This land shall be in perturbation.
> I show thee a land upside down: that occurred which never yet had occurred before.

. . . and in the Ipuwer papyrus in the Leiden museum:

> Plague is throughout the land.
> Blood is everywhere. The river is blood. Men shrink from tasting . . . and thirst after water.
> All is ruin! Forsooth gates, columns and walls are consumed by fire. The towns are destroyed. Oh, that the Earth would cease from noise, and tumult be no more.
> Trees are destroyed. No fruit or herbs are found . . . hunger . . . grain has perished on every side.
> All animals, their hearts weep . . . cattle moan. Behold, cattle are left to stray, and there is none to gather them together.
> The land is without light.

. . . and finally, an inscription on a shrine at El Arish, dating from Ptolemaic or Hellenistic times, tells of events that happened in the reign of a king Thom (or Thoum):

> The land was in great affliction. Evil fell on this Earth. . . .
> It was a great upheaval of the residence. . . .
> Nobody left the palace during nine days, and during these nine days of upheaval there was such a tempest that neither the men nor the gods could see the faces of their next.

Without doubt, something completely unprecedented seems to have happened, including phenomena which could be direct references to an ash-cloud dimout and blackout, and to sound and shock waves from the major paroxysms of the Bronze Age eruption. There is also some implication of an earthquake.

No sooner were these fragmentary Egyptian writings deciphered than their similarities to the biblical descriptions of the plagues of Egypt were remarked. It must be admitted that some of the parallels are closer when short excerpts are compared out of context, but they nevertheless are too startlingly alike to be dismissed as mere coincidence. Recently J. G. Bennett [12] pointed out that both the Egyptian and the biblical accounts have many features in common with reports of the Tambora eruption of 1815, as well as the better documented Krakatau eruption of 1883. In his account of the Tambora disaster in his *History of Java*, Sir Stamford Raffles mentions obscuring of the sun, a rain of stones and hail, plagues of insects, a whirlwind, destruction of crops (by ashfall) and of animals (for lack of forage), and epidemics—all attributable in one way or another to the effects of the eruption. Bennett proposed that the plagues of Egypt which paved the way for the Exodus actually were effects of the Santorin eruption as they would have been felt a few hundred miles away.

Biblical scholars do not yet agree on the probable date of the Exodus, but one of the possibilities is 1446-1447 B.C. (the other is about 1200 B.C.). This date is based on the statement in I Kings 6:1 that Solomon began to build the temple of Jerusalem in the fourth year of his reign, 480 years after the Exodus; Solomon's reign has been fixed with a fair degree of certainty as 970-930 B.C. Let us assume for now that 1447 B.C. is approximately the correct date, and examine the plagues of Egypt in the light of their possible connection with the Minoan eruption.

1. "... and all the waters that were in the river turned to blood. And the fish that was in the river died, and the river stank, and the Egyptians could not drink of the water of the river; and there was blood throughout all the land of Egypt" (Exodus 7:20-21).

Galanopoulos [67, 69, 75], elaborating on Bennett's suggestion, has suggested that rains laden with pinkish ash could have fallen on Egypt. It is true that the lowest Minoan tephra on Santorin is pale pink in color, and that some fine ash from that phase of the eruption might have reached Egypt; however, most of what was carried that far would have been the whitish upper pumice. Both the biblical and the Egyptian accounts lay heavy stress on the sanguinary aspect of the waters, but even large quantities of pinkish ash would hardly suggest blood, even figuratively. And if reddish or pinkish material had come down in a rainstorm in such large quantities that the *river* waters were noticeably tinged, would not the rain itself have been even more colorful and more likely to have inspired a report of a *rain* of blood?

There is an easier way to turn waters red as blood. Many instances of red water have been recorded throughout history. Both fresh and salt water algae produce red pigments, and so do some protozoa. In his textbook *Modern Microbiology*, W. Umbreit [249:294] states: "None [of the algae] cause disease but some, on growth, produce poisonous substances. Sometimes they grow so vigorously as to produce poisonous tides, which may be green, red, yellow, or brown. As a matter of interest, this kind of growth of algae may well have been one of the plagues visited upon the Egyptians. . . ."

Not only could a proliferation of algae or other pigment-producing organisms have turned all the waters in the river and ponds into "blood," but there is a faint chance that such a proliferation could have had some connection with the Santorin eruption. In 1850, C. G. Ehrenberg [54] documented more than fifty instances of bloody rain and snow from 730 B.C. to A.D. 1850, and gave a list of various algae responsible for specific instances. Many of these occurrences apparently coincided with meteor showers and other abnormal phenomena. It would be highly interesting to explore the possibility that the normal meteorological regime of the whole eastern Mediterranean was sufficiently disturbed by the Santorin eruption that it provided conditions favorable for the growth of some pigment-producing organism in Egypt. But that is a task for some "biomythologist" [79].

2. ". . . and the frogs came up, and covered the land of Egypt" (Exodus 8:6).

Galanopoulos suggests that meteorological disturbances caused by the eruption gave rise to whirlwinds, which, passing over lakes or rivers, snatched up frogs and rained them down elsewhere. Cases where it has literally rained frogs and other small animals are not unknown. But again, I believe there is a much simpler explanation for the frogs.

One summer in New Mexico I learned at first hand how abnormal dampness in the desert can literally bring frogs out of the ground by the thousands. After a cloudburst miles away in the mountains one day, a flash flood roared down a normally dry arroyo and overflowed onto a stretch of level ground. There the water stood long enough to soak in, and suddenly there were frogs everywhere! They evidently had been in the ground in a state of suspended animation, awaiting moisture to emerge and feed. They were not quite as ubiquitous as the biblical frogs, but they definitely were a plague with their incessant chorus, night and day. In a few days the water dried up and they vanished as suddenly as they had appeared.

Could unusually heavy rains, a remote and indirect result of the Santorin eruption, have been responsible for the plague of frogs in some such way?

3. ". . . and the dust of the earth . . . became lice in man and beast; all the dust of the land became lice throughout all the land of Egypt" (Exodus 8:17).

4. ". . . and there came a grievous swarm of flies into the house of Pharaoh, and into his servants' houses, and into all the land of Egypt: the land was corrupted by reason of the swarm of flies" (Exodus 8:24).

If abnormally damp conditions brought forth the frogs, they could also have favored the breeding of lice and flies, which would take care of plagues three and four.

5. ". . . and all the cattle of Egypt died but of the cattle of the children of Israel died not one" (Exodus 9:6).

6. ". . . And they took the ashes of the furnace, and stood before Pharaoh; and Moses sprinkled it up toward heaven; and it became a boil breaking forth with blains upon man, and upon beast'" (Exodus 9:10).

The "murrain of beasts" is decidedly reminiscent of the poisoning of Icelandic sheep in the 1947 [238:22] and 1970 [239:46-48] eruptions of Hekla. During the first two hours of the relatively minor eruption which began on May 5, 1970, violent explosions showered ash over about twenty thousand square kilometers to the north and northwest of the volcano. The coarser particles, which fell near the volcano, contained a hundred parts per million of fluorine; the finer particles, which fell lightly on some of the country's most important grazing lands, contained two thousand parts per million. Sheep which ate grass dusted with the poisonous ash either died within a few days of acute effects, if their dose was high enough, or died later of starvation when chronic effects prevented them from grazing. Fortunately most farmers heeded official warnings and kept their cows, and as many sheep as possible, indoors until the rains could wash the harmful dust from the grass. Nevertheless at least 6,000 lambs and 1,500 ewes were lost.

The highly toxic ash from this relatively minor eruption was carried more than two hundred kilometers from its source; how much farther is unknown, as it fell out to sea. In most places it was only a little more than a millimeter thick on the average. It is not unreasonable to speculate that the Minoan ash, also produced from a rather silica-rich magma, might also have contained fluorine or some such substance capable of poisoning livestock. In that case, of course, there also should have been traces of fluorine, or whatever, in the ash which fell on Crete; but there the concentrations might not have been so high as to compound the damage caused by the ashfall *per se*. Only the finest ash would have reached Egypt, and in the case of Hekla it was the finest ash that carried the highest concentrations of fluorine.

Of course it is hardly likely that this, or any other misfortune, could have affected the Egyptians' cattle and at the same time spared the Israelites', so we must put this detail down to a chauvinistic exaggeration on the part of the tradition bearer—unless for some reason the Israelites happened to have sheltered and hand-fed their cattle during the crucial time.

Moreover, it is not unreasonable to speculate that people unaccustomed to bathing, and animals likewise, might break out in "boils and blains" if dusted with ash containing such an irritant. The implication of ash falling from heaven is a bit suggestive too, even though it is said to have been first tossed skyward by Moses.

7. "And Moses stretched forth his rod toward heaven: and the

Lord sent thunder and hail, and the fire ran along upon the ground. . . . So there was hail, and fire mingled with the hail, very grievous, such as there was none like it in all the land of Egypt since it became a nation. And the hail smote throughout all the land of Egypt all that was in the field, both man and beast; and the hail smote every herb of the field and brake every tree of the field" (Exodus 10: 23-26).

Hail is very commonly associated with an ash fall, for the ash particles act as nuclei around which ice can condense. Furthermore, all sorts of electrical phenomena may be associated with an ash fall, even quite far away from the eruption. At the time of the 1766 eruption of Hekla, for instance, the air was so heavily charged with electricity at Skagafjördur, some two hundred miles from the volcano, that what looked like tongues of flame were seen rising from iron weathervanes on a church roof, and at Holar in the same vicinity, poles and alpenstocks glimmered with a phosphorescent light [237]. Phosphorescent fallout was also observed fifty miles from the Krakatau eruption [108:30-31].

8. "And the Lord brought an east wind upon the land all day, and all that night; and when it was morning, the east wind brought locusts . . . and there remained not any green thing . . . throughout the land of Egypt" (Exodus 10:13, 15).

The plague of locusts is difficult to relate to the Minoan eruption in any way, unless the east wind which brought them was also part of meteorological effects. Otherwise the locusts were just a coincidence, adding insult to injury. Disastrous swarms of locusts, of course, are by no means rare in that part of the world.

9. "And there was thick darkness in all the land of Egypt three days: They saw not one another, neither rose any from his place for three days . . ." (Exodus 10:22).

What better explanation could there be than the thick darkness which is certain to have occurred as a result of the Krakatau-like eruption of Santorin? The discrepancy between the Egyptian and biblical versions as to the duration of the blackout can easily be attributed to the fact that both chronicles were compiled hundreds of years after the event.

10. "And it came to pass that at midnight the Lord smote all

the firstborn of the land of Egypt . . . and all the firstborn of cattle"
(Exodus 12:29).

Widespread sickness and death among humans as well as among
cattle could have resulted from indirect effects of the eruption, for
lice and flies bred by abormally damp conditions could have carried
disease. Although in historical eruptions of Santorin, noxious gases
have caused people to become nauseated or even more violently
upset as much as sixty miles downwind, it is not likely that any such
direct emanations could have reached Egypt, and certainly not in
doses concentrated enough to kill even the very young or very
infirm. In any event, the selectivity which took only the firstborn
can be only taken figuratively.

All in all, the plague of darkness is the strongest link to the
Santorin eruption; indeed, it is difficult to explain it in any way
other than as the effect of an ash cloud. All the other possible links
are purely speculative. They constitute an admittedly shaky in-
verted pyramid of *ifs: if* the eruption influenced the weather over
a wide area (as it well might have, at least to the extent of bringing
on more rain than normal) and *if* as a result there were thunderstorms and hail and enough rain to make things damp in a normally
very dry land, then conditions resembling some of the plagues
might have been produced. The presence of fluorine in the ash can
neither be proved nor disproved, and therefore remains only a possibility. But whenever it occurred with respect to Egyptian history,
the Minoan eruption must have been responsible for a series of unusual, largely unpleasant, and even highly alarming manifestations
there, which would have seemed to be of supernatural origin. If
those manifestations did come when Moses was agitating for the
Israelites' release from bondage, they could well have been the reason Pharaoh was finally persuaded to let the people go.

Also consistent with the idea that Exodus may have been influenced by the Minoan eruption of Santorin is the fact that some
of the early church fathers, such as Augustine, Eusebius, and Isidore, Archbishop of Seville, attempted to correlate Deukalion and
Ogyges of the Greek flood traditions with biblical figures [253:
159-160; 63:159]. Isidore made Ogyges the contemporary of Jacob,
and Deukalion the contemporary of Moses. Julius Africanus, however, considered Ogyges to be Moses' contemporary [63:159]. All

of them accepted the well established Greek tradition that Ogyges' flood was the earlier. Velikovsky, whose cosmic collision theory requires two catastrophes, but the lesser after the greater, correlates Deukalion with the Exodus, and moves Ogyges' flood up to the time of Joshua [253:160].

The most dramatic part of the story of Exodus is the miracle of the parting of the Red Sea waters. Here again the similarity between the Bible and the El Arish inscription leaves no room for doubt that something very abnormal happened, with consequences disastrous to an Egyptian king. The inscription tells that during the time of darkness and tempest, king Thom (or Thoum) led his forces against the "companions of Apopi" (the Egyptian god of Darkness), and was lost in a whirlpool at a place named Pi-Kharoti. This is unmistakably the same place as Pi-ha-Hiroth (or -Khiroth), where the Israelites camped just before the crossing. Unfortunately it is not known where Pi-ha-Hiroth was. Nor does the Bible make clear which Pharaoh ruled at the time of Exodus. His name is given as "Rameses," which many scholars believe means Raamses III, who ruled in about 1200 B.C. If Exodus took place 480 years before Solomon started building his temple, however, then the king in question could not have been Raamses III or any other Raamses, for according to presently accepted chronology it was Tuthmosis III who ruled at the time. The case for Tuthmosis III as the pharaoh of Exodus seems to be consistent with Egyptian sources: the Egyptian historian Manetho refers to a king named "Tutimaeus" or "Timaios," in whose days a blast of heavenly displeasure fell upon Egypt. (Could "Thom" or "Thuom" be yet another version of the name Tuthmosis? Manetho's work survives only in the form of citations by later writers.)

It has been argued that Tuthmosis III was far too strong a ruler to have permitted the Israelites to leave Egypt and invade Canaan, a land controlled by Egypt at the time, whereas under the weaker rule of Raamses III Egypt was beginning to disintegrate. This argument has been countered by Bennett [12], who points out that the Book of Exodus definitely depicts Pharaoh as a powerful autocrat, a ruler who could be swayed only by some terror before which he was helpless.

At any rate, Galanopoulos [67] has suggested that not only the

plagues of Egypt but also the miracle of the Red Sea crossing can
be explained in terms of the Santorin eruption—in this case, by the
tsunami resulting from the caldera collapse which could have been
responsible for Deukalion's flood. For this, Pi-ha-Hiroth would
have to have been on the shores of the Mediterranean. Biblical
scholars have long argued that the body of water crossed by the
Israelites was not the Red Sea we know by that name, but a body
of brackish water. The Hebrew name for it in the original manuscript is "Jam Suf," which means "Reed Sea," and reeds do not
grow in salt water. On the coast of the Sinai peninsula, along what
has long been the main route from Egypt to the countries of the
eastern Mediterranean, there is a lagoon now called the Sebkha el
Bardawil, known in Herodotus' time as the Sirbonis Lake (Fig. 38).
This is the choice of some scholars as the site of the crossing, and

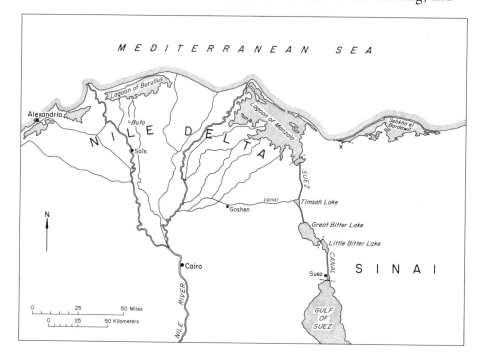

FIGURE 38. *Map of the Nile Delta and adjacent areas. The distributaries
of the Nile and the Sebkha el Bardawil are shown as they look today;
they probably were somewhat different at the time of the Exodus. Dotted
lines show various places that have been proposed as the real site of the
Red Sea crossing by the Israelites.*

it is also the choice of Galanopoulos. Offshore bars and spits, inter-rupted by narrow channels, separate the waters of such a lagoon from the open sea. During the withdrawal of the waters preceding a tsunami the lagoon coud be partially or wholly emptied, while the returning waters would flood over the bars in a rush. The Israelites thus could have crossed an exposed stretch of lagoon bot-tom during the retreat of the waters, while their pursuers were caught by the tsunami itself.

There can be no question that the effects of any large tsunami, particularly an unusually large one generated by caldera collapse, would have been much as pictured at the Sebkha el Bardawil and other lagoons along the coast. The main difficulty with the hy-pothesis is the matter of timing. On the face of things, the co-incidence that the Israelites happened to be at precisely the right spot at precisely the right time appears to be of such astronomical proportions that it would require divine intervention to bring it off, and thus in effect another miracle is being substituted for the one for which a scientific explanation is being sought. On second thought, however, the coincidence is a little less staggering. If it is assumed that the Minoan eruption was the underlying cause of at least some of the plagues which paved the way for the Exodus, and if the collapse of a substantial part of the caldera occurred shortly after the climax of that eruption, then the Israelites could have been at the coast at the opportune time, provided that their route did in fact lay that way. Moreover, the Sebkha el Bardawil is not the only site for which the theory is valid. The proposed mechanism would have operated equally well at the Lagoon of Manzala (see Fig. 38), another site which scholars have considered possible.

But even if we can place the Israelites on the spot at the right time, it is hard to imagine how the crossing itself could have been accomplished in the time available between the retreat and return of the waters. Usually that interval lasts up to half an hour at most. For the exceptional case of a Santorin tsunami, that might be stretched a bit, but any gain in time is offset by the fact that the waters of a lagoon, having only narrow channels through which to drain out, cannot ebb as rapidly as from an open coast even though the water is relatively shallow; the returning wave, how-ever, is not thus restricted, since it can pour over the sandy strips separating the lagoon from the sea. Thus the interval between maximum exposure of dry bottom and maximum flooding is shorter than on an open coast. Even if the Children of Israel numbered only about six hundred at the time, as Ben Gurion has suggested, it must have been quite a scramble for that many men, women, and children, accompanied by livestock and laden with all the possessions necessary for survival in the wilderness, to have accomplished the crossing of what possibly was a very muddy expanse within the limited time at their disposal. Nevertheless it is not impossible, particularly if they had to cross only a small corner of the lagoon. The spot marked X on Figure 38, for instance, would have been one of the first places to be exposed by the retreating waters, whereas the channel joining the lagoon with the sea (the spot suggested by Galanopoulos and Bacon [78: 197]) would have been the last place to go dry.

From the present configuration of the Sebkha el Bardawil (if we accept that as the site, for the moment), it is impossible to determine where the Israelites could have camped or exactly where they could have crossed, for the sandbars and spits framing lagoons are variable topographic features, constantly being modified by the action of waves and currents. A map of the Sebkha el Bardawil published in 1875 shows a quite different shape, with two channels communicating with the sea. Strabo reports that there was but one gap. Who can say what it was like in about 1447 B.C.? However, possible differences in detail do not in any way invalidate the basic premises of this mechanism of the crossing, and if it should ever be proved that Pi-ha-Hiroth was on the coast, Galanopoulos' theory would be tremendously strengthened. Until that time, however, other sites which have been proposed cannot be neglected.

The biblical description of the site is tantalizingly uninformative as to the location. The Israelites were instructed to "turn and encamp before Pi-Hahiroth, between Migdol and the sea, over against Baalzephon; before it shall ye camp by the sea" (Exodus 14:2). Of the crossing itself it says: ". . . and the Lord caused the sea to go back by a strong wind all that night and made the sea dry land, and the waters were divided" (Exodus 14:21). Egyptian records thought to refer to this epoch also mention tempests. High winds could drive back the relatively shallow waters at the head of the Gulf of Aqaba or the Gulf of Suez, or in one of the lakes now joined together by the Suez Canal (see Fig. 38). This would have given the Israelites as much as several hours in which to cut across, rather than go around, the water body before which they were camped; then a sudden drop in the wind, or change in its direction, could have allowed the waters to return in a rush to drown the Egyptians. The head of the Gulf of Suez or Gulf of Aqaba may be too salty to have been the "Reed Sea"; the Bitter Lakes would seem to be a better choice. From the shape of their shoreline (see Fig. 38) it is not difficult to conceive that a strong wind from the east might drive the waters of the Great Bitter Lake basin toward its north-western end, exposing the shallow ridge which separates it from the Little Bitter Lake basin (in which there is much less water, and in which whatever pile-up did occur would have been against the western shore and not against the ridge separating the basins). A Bitter Lake site, proposed some time ago by Sir William Dawson, is to my mind more in keeping with the idea that the waters were divided, as stated in the Bible.

There are at least three passages in the Bible, other than those in Exodus, which specifically allude to "Caphtor," as Crete was known to the Hebrews. One mentions the fact that the Philistines emigrated from Crete, and the others graphically describe a catastrophe which overwhelmed the "land of the Philistines" in terms which unmistakably suggest two of the main consequences of the Minoan eruption—darkness and tsunami. Amos 9:5-7 (written in the ninth century B.C.) says:

> And the Lord God of hosts is he that toucheth the land, and it shall melt, and all that dwell therein shall mourn: and it shall rise up wholly like a flood; and shall be drowned, as by the flood of Egypt. It is he . . . that calleth for the waters of the sea, and poureth them out upon the face of the earth. . . . Have I not brought up Israel out of the land of Egypt, and the Philistines from Caphtor, and the Syrians from Kir?

This has been construed to mean that the Exodus and the emigration of the Philistines were contemporaneous, but it could equally well refer to events that happened at different times. (Nobody seems to know much about the "Syrians from Kir.") Zephaniah 1:15, 17 and 2:5 (dating from the seventh century B.C.) tells us that:

That day is a day of wrath, a day of trouble and distress, a day of darkness and gloominess, a day of cloud and thick darkness. . . . And I will bring distress upon men, that they shall walk like blind men. . . . Woe unto the inhabitants of the sea-coast . . . the land of the Philistines. I will destroy even thee, that there shall be no inhabitants.

Jeremiah 47:2, 4, written in the sixth century B.C., refers even more specifically to a tsunami:

Behold, waters rise up out of the north, and shall be an overflowing flood, and shall overflow the land. . . . Because of the day that cometh to spoil all the Philistines . . . for the Lord will spoil the Philistines, the remnant of the country of Caphtor.

This passage seems to imply that the disaster overtook the "remnant of the country of Caphtor" *after* they had established themselves in Philistia (which they did in about 1200 B.C.), and standing alone could refer to some tsunami later than one caused by the Santorin collapse; however, the cities of Philistia were established well back from the coast where they were safe from the sea. Moreover, "waters rise up out of the north" is a more apt description of the approach of a tsunami to Crete, not to Philistia, whose coastline runs roughly north-south. The "land of the Philistines" in the passage from Zephaniah, of course, could mean Caphtor, whence they came, and it is highly descriptive of conditions that must have prevailed in Crete during the eruption and at the time of a tsunami. Any minor discrepancies are easily resolved if one remembers that the passages quoted were written so long after the event (even though in the form of prophecy) that if they are traditions concerning the Minoan eruption, the dispersal of the Minoans to other lands, and the settlement of Philistia, all those things could have seemed to later chroniclers to be contemporaneous even though they really were spread out over more than two centuries.

Whether the Philistines came directly from Crete in 1200 B.C. is not certain. Recent archeological findings suggest that they wandered for perhaps a generation before finding a place to settle. They were one of several tribes (not necessarily related, but all displaced) of the "Peoples of the Sea" (or, more properly, "people from beyond the sea"), who all migrated to the eastern Mediterranean region at about the same time. New excavations on long-neglected Philistine cities have completely revised former ideas about them

[9]. They were not Semitic, as previously thought. Whatever language they spoke, it was not Hebrew, but what it was is not yet known. One suggestion is Luvian, which, if proved true, would lend support to the idea that Minoan Linear A may have been Luvian. It has also emerged that the Philistines were not the uncouth boors pictured in the Bible and by the Egyptians. Excavations at Ashdod show that they even knew such refinements as bathtubs. But then, their relationships with the Israelites can hardly be termed cordial, and they tried to wrest territory from the Egyptians before establishing themselves in Philistia; so one can hardly expect that they would have been regarded with sympathy by either the Israelites or the Egyptians.

There is some confusion concerning the name Caphtor. According to the *Encyclopaedia Britannica* it "is to be identified with the Egyptian Ka(p)tar, which in later Ptolemaic times seems to mean Phoenicia, although the earlier *Keftiu* denoted Crete." Someone has suggested that the Minoans may have been the ancestors—spiritual if not physical—of the Phoenicians; but the Phoenicians were a Semitic people whose history goes back to at least 1600 B.C. After the collapse of Minoan Crete it is more than likely that some Minoan emigrés were assimilated in the melting pot of Phoenicia; perhaps it was they who imparted to the Phoenicians what the *Britannica* calls their "strangely un-Semitic love of the sea." In any event it was the Phoenicians who, beginning in about 1200 B.C., emerged as the same kind of seafaring nation the Minoans had been before 1450 B.C. On the other hand the Philistines, who are known to have come from Crete even though the how and why are not completely clear, shunned the coast and established their cities safely inland. Do they represent a later migration (during the very unsettled times of about 1200 B.C.) of a Minoan remnant who—though their culture had become overlaid first with the customs of their Mycenean rulers and then with those of the lands through which they wandered before reaching Philistia—still retained the memory of a terrible disaster from the sea? The answer to that must be sought in archeology, not geology.

So far, we have looked into the possible relationship of the Bronze Age eruption of Santorin to Deukalion's deluge, to other local Greek traditions of inundation, to the demise of Minoan civilization, and to the Exodus. There are still other myths and semi-historical traditions which might also be traced back to that event. And why not? It surely was the most stupendous display of Nature's forces ever to have awed the Mediterranean world.

One such myth is that of Phaëthon. Phaëthon was the son of the nymph Clymene. Taunted by his playmates for not having a father, he was assured by his mother that his father was none other than Helios (Apollo), the Sun god himself. Phaëthon thereupon set out to find his father and be acknowledged by him. After a long journey eastward he reached the palace of the Sun, where he was received and treated royally. When his father promised him any gift he would name, Phaëthon rashly demanded the privilege of driving the chariot of the Sun across the heavens for one day. Dismayed, because he knew the boy's strength was not equal to the task, Helios tried to dissuade him, promising anything else instead. All in vain; the stubborn lad insisted, and the god had to keep his word. As expected, Phaëthon was quite unable to control the fiery steeds, and they plunged wildly from their appointed course. Wherever the chariot approached too close to the Earth, springs and rivers dried up—even the Nile fled and hid its head—and everything was scorched. Wherever the chariot was pulled too far away, the earth became covered with ice and snow. Earth called upon Zeus for help, and Zeus struck Phaëthon from the chariot with a thunderbolt. He fell to earth on the bank of the Eridanus (the Po) and was buried there by his weeping sisters, who turned into trees exuding amber.

With the possible exception of the *fimbulvetr* (see Chapter 4), Phaëthon is the only myth, so far as I have been able to ascertain, that has been mentioned in connection with very slow changes in man's environment. In the *Timaeus*, in the discussion preceding the story of Atlantis, Plato has the Egyptian priest tell Solon that the Phaëthon myth "really signifies a declination of the bodies moving around the earth and heavens, and a great conflagration of things upon the earth recurring at long intervals of time. . . ." On this basis it has been suggested that the myth might be attributed to the changes in climate which caused the several advances and retreats of continental glaciers in Pleistocene time and the general warming up since the last glaciation. But as Galanopoulos very reasonably points out [74], the average change in temperature of the Atlantic

Ocean since the end of the Pleistocene has been only about one degree centigrade per thousand years; that is enough to melt ice-caps over the centuries, but far too slight a change to be noticed by man. If the myth has any basis in fact, a catastrophic event is a far more likely source.

Galanopoulos has suggested that "flashing arcs" [74] or lightning [78:73] seen during the eruption, together with the drop in temperature associated with the presence of ash in the atmosphere, may account for Phaëthon. Flashing arcs, a unique volcanic phenomenon originating in the crater and spreading out in all directions with the velocity of sound, are due to sound waves whose spherical fronts of compression and dilatation refract light to different extents and thus become visible. (The same phenomenon can be observed from behind a cannon.) Spectacular lightning displays are also common close to erupting volcanoes, particularly submarine volcanoes [15].

While I fully agree that Phaëthon could be an outgrowth of the Santorin eruption, I think there is a more logical connection. It is well known that after the Krakatau eruption of 1883 volcanic dust remained suspended in the high atmosphere for a long time, causing spectacularly flaming sunsets in various parts of the world for some time after. So unusually brilliant was the display on October 30, 1883, that fire brigades were called out in two American cities (Poughkeepsie, New York, and New Haven, Connecticut) in the belief that fires were raging just to the west [25:49]. Such sunsets (which were particularly noticeable in dry periods) would have been seen far more widely than any other effect of the eruption, and I believe it was one of these which may have been chiefly responsible for the Phaëthon myth [263].* Suppose some individual or group of individuals, concerned about crops drying up in a drought some time after the eruption (or still smothered in ash, if they were on Crete), had been treated to the sight of such an extraordinary sunset. Might it not have suggested to them that the whole world was burning up? Moreover, sunsets of this type would have been observed not too long after a memorable blackout or dimout, with its concomitant drop in temperature. To the people

* Galanopoulos and Bacon [78:115] mention the fire brigade incidents, but not specifically in connection with Phaëthon.

of those days, a departure of the sun from its normal course might well have seemed the logical explanation for both these abnormal manifestations. An explanation of the fiery chariot itself in terms of lightning or flashing arcs witnessed at the time of the eruption [74] (mainly visible close to the volcano) is not essential, for the idea of the chariot following a prescribed course through the heavens could have antedated Phaëthon, being an explanation of the normal daily course of the sun. It is the departure from normality, not the existence of the sun's chariot *per se*, which would have demanded an explanation in terms of some disruption of the sun god's usual routine.

A legend very similar to the Phaëthon myth has been recorded among the Kwakiutl Indians of British Columbia [99]. The son of the Sun once climbed up to heaven, where his father, taking advantage of this opportunity to take a nap, entrusted the boy with the brilliant nose and ear jewels which gave light to the world and asked him to carry them across the sky for one day in his place. The lad was warned not to stray too near the earth lest he set it afire. All went well until midday, but then the boy became impatient and began to run, taking a shortcut. All would have burned up had not the Sun waked in time to see what was happening. He hastened to overtake his son, tore the jewels from his grasp, and hurled him into the sea. E. S. Hartland [99] used this tale to illustrate the "oneness of human nature"; as in the case of flood traditions, the similarity of this legend to Phaëthon could reflect a similar response to the same kind of stimulus. Flaming sunsets are common over more limited areas after forest fires or small eruptions, and all over the world after major eruptions; it is even conceivable that the Indian legend too could have been inspired by sunset glows caused by the Minoan eruption. But then there is always the possibility that Phaëthon could have been transplanted to British Columbia in the same way that Beowulf came to South Dakota (see Chapter 7), undergoing appropriate modifications to conform to some existing tradition of solar jewels. Who knows what goldseeker, for instance, might have swapped tales with the Indians some time before 1895, when the legend was first recorded? The names of many old mining camps throughout the West are ample testimony that gold rushes attracted a goodly number of people who had the benefit of a classical education. However, Galanopoulos and Bacon [78:73] mention that there is a Phaëthon-like myth from Guatemala; thus a worldwide phenomenon of some sort may be at the root of all three, and nothing could be more worldwide than atmospheric dust following a Krakatau-like eruption.

Another classical myth includes a detail which may embody a memory of the Minoan eruption. When Zeus tricked the faithful Alkmena into bed with him by assuming the form of her husband Amphitryon (thus begetting Hercules), he caused the night to last for the space of three days in order to escape the eagle-eye of his wife Hera (and perhaps to prolong his pleasure?). Marinatos believes this to be a reference to the extensive and no doubt long-lasting blackout which accompanied the climax of the eruption [156:252].

Luce's roundup of possible memories of the Santorin eruption in Greek literature includes various mentions of floating islands. The island of Delos in particular, it was said, originally floated around the Aegean and became fixed only after Apollo was born there. The floating islands could be the memory of the banks of pumice which, beyond any doubt, must have littered the sea after the eruption, like the banks of floating pumice sighted in various parts of the Indian Ocean after the Krakatau event [143:152-153].

The possibility has been suggested that the myth of Ikaros (Icarus) may be another repercussion of the Minoan eruption [22]. Ikaros was the son of Daedalus, that "most skilled artificer" who built the labyrinth at Knossos for King Minos. Daedalus lost favor with the king and was imprisoned in a tower, but he fashioned wings for himself and Ikaros out of feathers and wax and the two of them escaped by air. Ikaros ignored his father's warning not to fly too near the sun, whereupon the wax holding his wings melted and he plummeted into the sea near an island which thenceforth bore his name. A rock just off the southern shore of Ikaria is said to be Ikaros, turned to stone. The suggestion was that a volcanic bomb from Santorin may have inspired the myth, but if so, it could not have been observed to fall near Ikaria; even if the Minoan eruption was very much more powerful than that of Krakatau, no sizable ballistic fragments would have reached Ikaria, which is at the very limit of distribution of the fine airborne ash found in the deep-sea cores (see Fig. 33). The Ikaros rock matches the rock on shore, of which it presumably is an erosion remnant. Possibly the myth originated when someone observed the fall of a meteor into the sea [22]; possibly it is an imaginative invention. The name *Ikaria* is actually the Phoenician equivalent of the older name *Ichthyassa*, meaning "Island of Fish," but people living there today still believe that their rock is really the fallen Ikaros, and are even more convinced of it since the Greek government chose to erect a statue on it commemorating the birth of aviation [21].

Before leaving the eastern Mediterranean and its traditions which might be linked to Santorin, we should mention one other possible piece of evidence of the tsunami in that part of the world. Excavations at Ugarit, the ancient city near modern Latakia, have shown that its port and half the city were demolished in about 1400 B.C. A Phoenician poem found in the library of Ugarit speaks of tempest and tsunami, and has been taken to refer to that event. Ugarit survived the blow, whatever its cause, and continued until 1350 B.C., when it was finally destroyed by an earthquake. The date of this destructive wave is close enough to the time of the Minoan eruption, considering the uncertainty surrounding all the dates in question, that it just might have been a Santorin tsunami, rather than some earthquake tsunami.

Epilogue

THE EXAMPLES CITED IN THE FOREGOING PAGES have, I hope, demonstrated that geology and folklore and history are interrelated in a number of ways.

Geologic processes or their results have left their impression on folklore in the form of euhemeristic legends embodying the more or less distorted memory of real occurrences, usually catastrophic, and etiological myths reflecting man's innate desire to understand and explain his physical environment. The latter often do credit to the powers of observation of long-gone people, but rarely come anywhere close to the real reason for the phenomena they attempt to elucidate, and then only by coincidence.

However, before we are tempted to think patronizingly of the old philosophers or tribal wise men, let us remember two things: first, there are still among us many intelligent but uninformed people who believe equally impossible things concerning geologic phenomena, and some of these misconceptions are so widespread as to constitute a form of folklore in themselves; and second, occasionally some idea debunked as folklore has turned out to be scientific fact after all.

While geologic processes and their results have undeniably had much more effect on folklore than vice versa, the relationship has not been entirely one-sided. Geologic nomenclature is indebted to folklore for a number of terms, particularly in volcanology, and on rare occasions local legends have been of help in working out details of local geology.

Epilogue

Atlantis, surely the most controversial myth of all time, belongs in a category all its own. Inasmuch as it is not embodied in the oral traditions of any nation or culture, it is not true folklore at all. It would come close to qualifying as the most colossal piece of fake-lore ever invented, but for the fact that there is abundant reason to believe that Plato, knowingly or otherwise, incorporated into his literary creation some repercussions of the Bronze Age eruption of Santorin which lend Atlantis a measure of factual basis, at least indirectly. In this light it emerges as a fascinating hybrid which straddles the border between invention and legend.

The case for an Aegean site of Atlantis thus does not depend on the validity of the theory of the volcanic destruction of Minoan Crete. Even if the sudden eclipse of Minoan power was not simultaneous with the destruction of the colony on Santorin, it followed so uncomfortably close in time that it is not impossible that the eruption could have contributed to the ultimate demise of Minoan Crete. In my science-fiction reconstruction of possible events (in Chapter 8) I have played down the role of tsunamis resulting from the collapse of the Santorin caldera. That is because I believe that if the collapse had occurred at the climax of the eruption, and suddenly enough to generate a tsunami at least comparable to the Krakatau tsunami, then there should have been more noticeable signs of economic decline during Late Minoan I B. That does not mean that a tsunami of more modest proportions, but large enough to be remembered in the story of Deukalion's deluge, could not have occurred then or later. But to postpone a very *large-scale* collapse for a generation or more after the climax of the eruption is not sound geologic reasoning (although it probably cannot be ruled out as absolutely impossible); a gradual collapse, causing smaller or no tsunamis, is far more likely.

To be sure, the Krakatau eruption did not destroy the Dutch East Indies, as opponents of the theory of the volcanic destruction of Minoan Crete have always been quick to point out. But its ash blew mainly out to sea, and what little fell on land posed no threat to agriculture in that tropical climate; a high birth rate would soon make up for the villagers wiped out by the tsunami; and there was no earthquake. Two eruptions of lesser magnitude than Krakatau did have very serious adverse effects on entire nations, however: Iceland came very close to being abandoned as a result of the

Lakagígar eruption of 1783, and the state of Old Mataram in Java seems to have collapsed mainly as a result of a tephra fall. Certainly, no better explanation has yet been offered for the astonishingly rapid descent of the Minoans from the heights of power than a knockout blow from Nature—probably delivered not all at once, but in the form of a one-two or even a triple punch; the ash fall (which is a fact), possible tsunamis (not necessarily catastrophic, but destructive), and a possible major earthquake, one following hard on the heels of the other in a space of, say, twenty years, might well have collectively paved the way for invasion of Crete by the Myceneans.

In all attempts to find a geologic event at the root of a myth or legend, it must be kept in mind that the proffered suggestions are merely possibilities. In some cases there is the chance that the intent was metaphorical from the beginning. In others, some of the explanations may be mutually exclusive. For instance, if Deukalion's deluge and the Exodus are some day dated precisely and prove to have happened a hundred or more years apart, they obviously cannot both be the result of the Bronze Age eruption of Santorin; if and when their dates, and the absolute date of the eruption, are finally established, then the case for a euhemeristic relationship of one or both of these traditions to the eruption will either be substantially strengthened, or will collapse. Until then in these cases, and probably for all time in most others, we are left swimming in a sea of speculation. But is it not a delightful sport?

APPENDIX A

The Beringer Case

INASMUCH AS THE FACTS of the Beringer case are quite different from the story generally accepted by geologists themselves, it constitutes an example of a very special kind of geologic folklore, the "professional geomyth." Most students of elementary geology or paleontology have heard the tragicomic tale of poor Professor Beringer, an avid collector and describer of fossils in the days when the idea of special creation was unquestioned and the nature and purpose of these obviously organic-looking forms recovered from solid rock were matters of lively speculation. It has been told and retold how Beringer's students decided to play a trick on him and manufactured outlandish forms which they "planted" for him to find. Not until after he had published a tome describing and illustrating them did he learn the truth —it is said he did not tumble to the deception until his own name was dug up—whereupon he ruined himself financially trying to track down and buy back every copy, and died heartbroken shortly after.

The true story is not so sad. Dr. Johann Bartholomew Adam Beringer (1667-1740) was not a professor, but a Würzburg physician and savant interested in the subject of "oryctics" ("things dug up from the earth"). He does indeed owe his place in the history of geology to a hoax perpetrated at his expense. According to M. E. Jahn and D. J. Woolf [122], who have translated and annotated his works, the motivation was more sinister than a mere student prank. Two colleagues conspired to discredit Beringer, apparently because they resented his undoubted arrogance and perhaps also because they envied his equally undoubted ability. At any rate, in May 1725 the three lads hired by Beringer as diggers (one of whom was in on the plot) came upon the first unusual find, carved likenesses of the sun and its rays and of worms. These were soon followed by perfect stone crabs, toads, flies and other insects, lizards, more worms, water snakes, complete flowering plants, and even Hebrew letters. When Beringer had plates engraved and was preparing a dissertation on the curious stones, the conspirators decided the joke had gone far enough and spread the rumor that the finds were false. Beringer would not believe it. They sold some of their "fossils" to the worthy doctor and then confessed to their manufacture, but he stubbornly clung to the opinion that they were trying to minimize his great discoveries out of jealousy, and went ahead with publication of his treatise. When he finally was forced to the realization that he had been duped—possibly by finding his own name petrified, as the legend has it—he was more furious than heartbroken, and instituted legal proceedings. The truth came to light in the questioning of his diggers, and it was the conspirators who were discredited and ruined, not he. He did not die until fourteen years later, during which time he published at least two books of some merit; but his name is remembered today only because of his "Lying Stones."

The Piltdown Hoax

The Piltdown hoax [267] was intended to fool scientists as a whole, and it did succeed in fooling some for about forty years. In 1912 a human brain case and an ape-like jaw were found, along with primitive artifacts and Pliocene mammal remains, in a gravel deposit at Piltdown, England. The find was named *Eoanthropus dawsoni* (Dawson's Dawn Man) in honor of the discoverer, Charles Dawson, a lawyer who was also a highly competent amateur geologist and archeologist. From the outset scientists were divided into two camps over Dawn Man. The "monists" were convinced that the cranium and jaw did in fact represent a remote ancestor of man in whom human and simian characteristics were combined—a veritable "missing link." The "dualists" claimed they must represent two creatures, a late Pleistocene human and a Pliocene ape, washed into the gravel from an older deposit together with the Pliocene mammal remains. The finding of the remains of a second individual about two miles away convinced all but the most skeptical that *Eoanthropus* was indeed a primitive hominid. Part of this second find, a tooth, was turned up by as unimpeachable a witness as the French anthropologist-philosopher Teilhard de Chardin, who had offered his help in the exciting search for further evidence. The precise agreement of this tooth, down to the last detail, with the shape predicted on the basis of the supposedly matching cranium and jaw was in itself too good to be true and should have been suspicious; but why should anyone have suspected that the Dawn Man was nothing more than an extremely elaborate and unusually well planned fraud?

As new discoveries of fossil man were made in other parts of the world, Piltdown Man began to look more and more anomalous geographically, geologically, and particularly anthropologically. Finally, in 1949, the fluorine test, developed by J. S. Weiner, was applied to the Piltdown bones. This test is based on the fact that fluorine present in soil water will gradually accumulate in buried bones and teeth; thus older bones should contain larger amounts than younger ones. The jaws and cranium of Piltdown Man proved to be of different ages, and both very low in fluorine—no older than fifty thousand years. This presented such a problem for both "monists" and "dualists" alike that the idea of a hoax finally and inevitably suggested itself as the only possible answer. Very painstaking investigations ultimately revealed that the cranium was indeed that of a fossil human, but not a very ancient one, and that the jaw was from a modern ape. The bones and tooth had been very cleverly stained and abraded to look as old as Pliocene, and planted along with appropriate mammal remains and flints. Circumstantial evidence points to Charles Dawson as the sole perpetrator of the forgery, but as he died long before the hoax was recognized, the complete facts may never be known. Some who had steadfastly refused to accept Piltdown Man also died before their position was vindicated, but many others lived to say "I told you so!"

Appendix B

Radioactive Dating Methods

As we have referred to radiocarbon dates several times, it might be well to include a brief explanation of radioactive dating methods in general and of the carbon-14 method in particular. All are based on the fact that radioactive elements decay (that is, they spontaneously give off nuclear particles, thereby producing new elements) at a constant, measurable rate called the half life, which is the time required for half the nuclei in a sample of the element to decay. The ratio of the amount of the radioactive element present at a given time to the amount of the end product of decay tells how long the decay process has been going on, or in other words, how old the sample is. Most of the radioactive elements present in rocks and used for geochronological purposes have very long half lives: uranium and thorium (which decay to lead), rubidium-87 (which decays to strontium-87), and potassium-40 (which ends up as argon-40) are the most widely used. A long half life is necessary in most geologic contexts, where we are dealing with rocks millions or billions of years old. However, the long-lived elements are of no use at all when we want to date relatively recent materials, and that is where W. F. Libby's radiocarbon [141] (short for radioactive carbon) method comes in handy.

A certain, very small proportion of carbon atoms have an atomic weight of 14 instead of the usual 12; they are formed by cosmic ray reactions with nitrogen in the upper atmosphere. Carbon-14 is unstable (radioactive) and spontaneously decays to nitrogen-14. "Heavy" carbon atoms combine with oxygen in the upper atmosphere to form "heavy" carbon dioxide, which eventually can be taken up by living things along with "normal" carbon dioxide. The distribution of carbon-14 is constant throughout the world at any given time, and all living things contain a small amount of it, which is constantly replenished by life processes as it decays. When the organism dies, the replenishment ceases and the radioactive carbon decay becomes measurable. If we compare the amount of carbon-14 remaining in ancient organic matter —such as peat, shells, buried wood, or charcoal from prehistoric hearths —with the universal abundance, the time elapsed since the organism died can be computed.

The greatest difficulty with radiocarbon dating is finding material that has been preserved from contamination. Percolating ground water, chemical alteration, or washing in of older material (in lake deposits) may change the proportion of radioactive carbon present in a sample and thus give an age which is not the true age. Recently it has been found that the rate of production of carbon-14 cosmic rays has not been constant, as formerly believed, but has varied slightly over the centuries [228] because the intensity of the earth's magnetic field has varied, with small, more or less cyclic fluctuations superposed on a larger, generally linear trend, which is itself just a part of a much longer cycle. When the geomagnetic field is weaker, the earth is less shielded from the cosmic rays which produce carbon-14 in the upper atmosphere. Compared with the errors that can be caused by contamination, those due to cosmic-ray variations are small, but they get larger the older the sample. Tables giving the exact corrections for the effect have been worked out for the past seven thousand years by calibrating radiocarbon ages against tree-ring ages, which can be dated to the exact year by dendrochronology.

Until recently radiocarbon dates were all computed using a carbon-14 half life value of 5,568 years. The half life has since been determined more accurately as 5,730 years, and dates now are calculated on the basis of this "preferred" value. However, dates will continue to be reported in terms of the old half life value as well. Radiocarbon dates are given as "—— years B.P." ("before present"), and if it is necessary to express them in terms of the Christian calendar, 1950 is taken as the reference year. Thus a date of "3,500 ± 100 years B.P." would correspond to 1600 B.C., give or take a hundred years. For geologic purposes the present precision of the radiocarbon method is adequate; to date a 10,000-year-old postglacial deposit to within ± 50 or ± 100 years is accurate enough for all practical purposes. But for archeologists, who may be interested in very fine subdivisions of time—a generation or so— within the past few millennia, radiocarbon dating is handy but not precise enough in its present stage of development. Results of very recent research, however, give every reason to hope that very fine determinations, permitting close dating in terms of the Christian calendar, will be possible in the very near future [184].

Appendix C

The Mammoth Deep-Freeze

Dear to the hearts of those whose theories demand some kind of global catastrophe to explain the prevalence of deluge traditions are the mammoths and other animals whose remains have been found frozen in the Arctic permafrost. The number of well-preserved animals has been quoted as being in the tens of thousands (mammoths alone) or in the millions (all kinds of animals) [127:226]. The general idea is that whole herds of animals, peacefully feeding on a balmy day, were annihilated in the catastrophe and their carcasses whisked to polar latitudes in a matter of hours as a result of a shift of the earth's axis of rotation or a slippage of its outer shell [127:224; 253:330].

The fact is, very few *whole* mammoths have been found in a good state of preservation—only four, to be exact [33:807]. More often, bits of hair and skin and even meat have been found along with dismembered skeletons, but by and large it is mainly bones and tusks which have been found. (The tusks, being most durable chemically, outlast everything else. In Siberia alone some fifty thousand mammoth tusks have been collected for the ivory trade over the past few centuries, and much of the ivory in use today is from that source.) Collectively, these remains represent animals that lived and died over a period of thousands of years [60:470].

The most famous and most fully documented instance of preservation of the entire body of a mammoth is the Berezovka mammoth, found in 1900 in eastern Siberia, where caving of a river bank had partially exposed it in frozen muck. A special expedition was sent to study it the following summer, by which time part of the exposed back had been eaten by wild animals. The presence of clotted blood in the chest, unswallowed food, and broken bones clearly indicated that the beast had met its death in a fall. Presumably it was feeding along the top of a river bluff one autumn day when the bank caved in under its weight, precipitating it into the muck of the flood plain below, where it soon froze [60:470]. If this occurred at the end of an interglacial stage, the body could have remained frozen until it was finally exposed by erosion. Another explanation which has been advanced for its complete preservation is that it fell into a tundra-covered frost crack in the permanently frozen ground, where it was preserved much as are the bodies of unfortunate climbers who fall into crevasses of Alpine glaciers. The latter, however, may be yielded up years later at the ice front, while the frozen mammoths had to wait for erosion or ivory hunters to uncover their remains.

One thing the catastrophists usually neglect to explain when using the frozen mammoths as evidence of a stupendous upheaval is, why have no human remains been found too? Any disaster which could wipe out tens of thousands of woolly mammoths at one fell swoop would have been equally hard on the cave men who hunted them. As for the arguments for an overnight shift into a colder climatic zone: if, prior to their demise, the mammoths were living in a mild climate, why were they all wearing such heavy wool coats? And why are the other animals whose remains are found with theirs also overwhelmingly of species which are at home in a cold climate, like the mastodon, moose, deer, bear, musk-ox, yak, bison, and wolf, to name a few?

Paleomagnetic evidence does indeed suggest that the magnetic poles of the earth have not always been in their present position but have wandered as though the whole earth has rolled over on its axis, or as though the outer shell of the earth has slipped relative to the interior (this is in addition to the drift of continents with respect to each other on the surface, as discussed in Chapter 9). But like most geologic processes, the polar wander has been accomplished exceedingly slowly in the course of time, at a rate imperceptible to living things. Likewise, the finding of the jumbled remains of many individuals in one deposit does not necessarily mean that they all perished at once, particularly when the remains are in different stages of decomposition; these concentrations could have accumulated over a long period in boggy spots, or as a result of redeposition by streams. The mammoths simply do not add up to a universal or a widespread catastrophe.

The frozen mammoths have also been the source of a misconception so widespread among geologists themselves that it constitutes a second example of professional folklore (the first being the Beringer case). The story is widely circulated that one of the Pleistocene mammoths found in Siberia was in such an excellent state of preservation that its meat was served at a banquet given in connection with a geological meeting in St. Petersburg. As I first heard the tale in my student days, the banquet was at an International Geological Congress; but in that case it could not have been the Berezovka mammoth, for it was not discovered until 1900 and the Seventh International Geological Congress met in St. Petersburg in 1897. Thus it would have to have been some other mammoth or some other meeting, and both possibilities are equally unlikely. Even the Berezovka mammoth, which was the best preserved, could hardly have been appetizing enough to serve to anyone after it was brought back to St. Petersburg, as a glance at the record will show.

O. F. Herz, the leader of the expedition sent to excavate the Berezovka mammoth in 1901, referred more than once in his report [109] to the terrible stench of the rotting meat: "Upon the left hind leg I also found portions of decayed flesh, in which the muscular bundles were easily discernible. The stench emitted by this extremity was unbearable, so that it was necessary to stop work every minute. A thorough washing failed to remove the horrible smell from our hands. . . ." And later: "Despite the fact that the mammoth is in a frozen condition, the stench emitted is very disagreeable"; and still later: "The stench is not near so intolerable as during the first two days, probably because we have grown accustomed to it."

When they got farther into the remains they did find some meat in a remarkably good state of preservation, however, and the following passage may be what gave rise to the fiction that it was fit for human consumption: "The flesh from under the shoulder . . . which is fibrous and marbled with fat, is dark red in color and looks as fresh as well-frozen beef or horsemeat. It looked so appetizing that we wondered for some time whether we should taste it, but no one would venture to take it into his mouth, and horseflesh was given the preference. The dogs cleaned up whatever mammoth meat was thrown to them."

It is remarkable enough that the meat was still fit for the dogs, but the fact remains that nobody could summon up the courage to try it when it was at its freshest. What would it have been like after the trip back to St. Petersburg, even in cold weather? It had taken the expedition four months to reach the site, and presumably as long to return. Nevertheless, the folklore of the mammoth meat banquet has even found its way into geology textbooks, and only sometimes with a cautious "it is said that. . . ."

REFERENCES

*The following is a partial list of works consulted
in the preparation of this book.*

1. Alsop, Joseph. *From the Silent Earth: A Report on the Greek Bronze Age.* New York: Harper and Row, 1962.

2. Andree, Richard. *Die Flutsagen, ethnographisch betrachtet.* Braunschweig: Friedrich Bieweg und Sohn, 1891.

3. Armstrong, Baxter H. "Acoustic emission prior to rockbursts and earthquakes." *Bulletin of the Seismological Society of America,* vol. 59 (1969), 1259-1279.

4. Askelsson, J., Bodvarsson, G., Einarsson, T., Kjartansson, G., and Thorarinsson, S. *On the Geology and Geophysics of Iceland.* International Geological Congress, 21st, Copenhagen, 1960, Guide to Excursion A-2.

5. Babcock, William. "Legendary islands of the Atlantic." *American Geographical Society Research Series,* no. 8 (1922), 11-33.

6. Baker, George. "The role of australites in aboriginal customs." *Memoirs of the National Museum of Victoria* (Melbourne), no. 22 (1957), pt. 8.

7. ———. "Tektites." *Memoirs of the National Museum of Victoria* (Melbourne), no. 23 (1959).

8. Baker, P. E., Gass, I. G., Harris, P. G., LeMaitre, R. W. "The volcanological report of the Royal Society Expedition to Tristan de Cunha, 1962." *Philosophical Transactions of the Royal Society of London,* series A, vol. 256, no. 1075 (1964).

9. Balmuth, Miriam. *The Philistines.* Lecture delivered at Indiana University, March 27, 1969.

10. Bemmelen, R[ein] W. van. "The influence of geologic events on human history (An example from Central Java)." *Koninklijk Nederlandsch Geologisch-Mijnbouwkundig Genootschap Verhandelingen,* geol. ser., Deel XXI (1956), 20-36.

11. ———. "Four volcanic outbursts that influenced human history. Toba, Sunda, Merapi and Thera." *Acta of the 1st International Scientific Congress on the Volcano of Thera, Greece, 1969.* Athens (1971), 5-50.

12. Bennett, J. G. "Geo-physics and human history: New light on Plato's Atlantis and the Exodus." *Systematics,* vol. 1 (1963), 127-156.

13. Berg, L. S. "Atlantis and Aegeis" ("Atlantida i Egeida"). *Priroda,* no. 4 (1928), 383-388.

14. Biel, Erwin R. "Climatology of the Mediterranean area." *Chicago University, Institute of Meteorology, Miscellaneous Reports,* no. 13 (1944).

15. Björnsson, Sveinbjörn, Blanchard, Duncan C., and Spencer, A. Theodore. "Charge generation due to contact of saline waters with molten lava." *Journal of Geophysical Research,* vol. 72 (1967), 1311-1323.

16. Blot, C. "Origin profonde des séismes superficiels et des éruptions

volcaniques." *Publications du Bureau Central Séismologique International* (Toulouse), ser. A, no. 23 (1964), 101-131.

17. Bretz, J. Harlen. "The Dalles type of river channel." *Journal of Geology*, vol. 32 (1924), 139-149.

18. Bright, John. "Has archeology found evidence of the Flood?" In *The Biblical Archeological Reader*, G. Ernest Wright and David Noel Freedman, eds. Chicago: Quadrangle Books, 1961.

19. British Naval Intelligence Geographic Handbooks. *Greece*, vols. 1-3, 1944.

20. Broecker, Wallace S. "Isotope geochemistry and the Pleistocene climatic record." In *The Quaternary of the United States*, H. E. Wright, Jr., and David G. Frey, eds. Princeton (New Jersey): Princeton University Press (1965), 737-753.

21. Brumbaugh, Ada. Personal communication (1970).

22. Brumbaugh, Robert S. "Plato's Atlantis." *Yale Alumni Magazine*, vol. 33 (1970), no. 5, 24-28.

23. Brunvand, J. H. *The Study of American Folklore: An Introduction.* New York: W. W. Norton, 1968.

24. Bulfinch, Thomas. *Bulfinch's Mythology.* Garden City (New York): Garden City Publishing Company, 1938.

25. Bullard, Fred M. *Volcanoes: in History, in Theory, in Eruption.* Austin: University of Texas Press, 1962.

26. Byerly, Perry. *Seismology.* New York: Prentice Hall, 1942.

27. ———. "The Fallon-Stillwater earthquakes of July 6, 1954 and August 23, 1954: Historical introduction." *Bulletin of the Seismological Society of America*, vol. 46 (1956), 1-3.

28. Cadogan, G., Harrison, R. K., and Strong, G. E. "Volcanic glass shards in Late Minoan I, Crete." *Antiquity*, vol. 46 (1972), no. 184, 310-313.

29. Careri, John Francis Gemelli. "A Voyage around the World. Containing the most remarkable Things He Saw in New Spain." *A Collection of Voyages and Travels*, vol. 4, London: Messrs. Churchill, 1732, 478-533.

30. Caskey, John L. "Crises in the Minoan-Mycenean world." *Proceedings of the American Philosophical Society*, vol. 113 (1969), 433-449.

31. Casson, Lionel. *Ancient Egypt.* New York: Time, Inc., 1965.

32. Chadwick, John. *The Decipherment of Linear B:* Cambridge (England): Cambridge University Press, 1958.

33. Charlesworth, J. K. *The Quaternary Era, with Special Reference to Its Glaciation.* London: Edward Arnold, 1957.

34. Clapp, Frederick G. "Geology and bitumens of the Dead Sea area, Palestine and Transjordan." *Bulletin of the American Association of Petroleum Geologists*, vol. 20 (1936), 881-909.

35. Clarke Blake. "America's greatest earthquake." *Reader's Digest*, vol. 94 (April 1969), 110-114.

36. Clark, Ella E. *Indian Legends of the Pacific Northwest.* Berkeley: University of California Press, 1953.

37. Clark, H. C., and Kennett, James P. "Confirmation of the reality of

the Laschamp geomagnetic polarity event in cores from the Gulf of Mexico." *EOS* (Transactions of the American Geophysical Union), vol. 53 (1972), 354.

38. Clark, R. H. "Volcanic activity on White Island, Bay of Plenty, 1966-1969. Part 1, Chronology and crater floor level changes." *New Zealand Journal of Geology and Geophysics*, vol. 13 (1970), 565-574.

39. Coombs, Howard A., and Howard, Arthur D. *Catalog of the Active Volcanoes of the World including Solfatara Fields. Part 9: United States of America.* Naples: International Volcanological Association, 1960.

40. Cowan, James. *Fairy Folk Tales of the Maori.* Auckland (New Zealand): Whitcomb and Tombs, 1925.

41. Davis, F. Hadland. *The Myths and Legends of Japan.* London: George C. Harrap and Co., 1912.

42. Davis, Simon. *The Decipherment of Minoan Linear A and Pictographic Scripts.* Johannesburg: Witwatersrand University Press, 1967.

43. Davison, Charles. *Great Earthquakes.* London: Thomas Murby and Co., 1936.

44. DeCamp, L. Sprague. *Lost Continents: The Atlantis Theme in History, Science, and Literature.* New York: Gnome Press, 1954; New York: Dover Publications, Inc., 1970.

45. DeLaguna, Frederica. "Geological confirmation of native traditions, Yakutat, Alaska." *American Antiquity*, vol. 23 (1958), 434.

46. Delendas, Peter A. *Guide of Santorin.* [Athens?], 1966.

47. Dewey, James, and Byerly, Perry. "The early history of seismometry (to 1900)." *Bulletin of the Seismological Society of America*, vol. 59 (1969), 183-227.

48. Dietz, Robert S. "Continent and ocean basin evolution by spreading of the sea floor." *Nature*, vol. 190 (1961), 854-857.

49. Dittmer, Wilhelm. *Te Tohunga: The Ancient Legends and Traditions of the Maoris.* London: Routledge and Sons, 1907.

50. Dolomieu, Déodat de. *Voyage aux îles de Lipari, Fait en 1781, ou Notices sur les îles Aéoliennes, pour Servir à l'Histoire des Volcans.* Paris: Rue et Hotel Serpente, 1783.

51. Donnelly, Ignatius. *Atlantis: The Antediluvian World.* New York: Harper and Bros., 1882. Modern revised edition, Egerton Sykes, ed. New York: Gramercy Publishing Co., 1949. (Page references are to the modern edition.)

52. Dorson, Richard M. "The debate over the trustworthiness of oral traditional history." *Volksüberlieferung*, Kurt Ranke Festschrift (1968), 19-35.

53. Duke, C. Martin. "The Chilean earthquakes of May 1960." *Science*, vol. 132 (1960), 1797-1802.

54. Ehrenberg, C. G. "Hr. Ehrenberg übergab eine reichlich Centurie historischer Nachträge zu den blutfarbigen und sogennanten Prodigien." 27. Juni Gesamtsitzung der Akademie. *Monatsberichte der Berlin Königl. Preuss. Akad. Wissenschaft* (1850), 215-246. [Cited in Gaughran, E.R.L.]

References

55. Einarsson, Thorleifur. "Geologie von Hellisheidi (Sudwest-Island)." *Sonderveröffentlichen des Geologischen Institutes des Universität Köln*, 5, 1960.

56. Ellis, William. *Polynesian Researches, During a Residence of Nearly Eight Years in the Society and Sandwich Islands*, vol. 4. New York: J. & J. Harper, 1833.

57. Emiliani, Cesare. "Pleistocene temperature." *Journal of Geology*, vol. 63 (1955), 538-578.

58. Evans, Arthur. *The Palace of Minos at Knossos*, vol. 2, part 1. London: Macmillan and Co., 1928.

59. Felton, Harold W. (ed.) *Legends of Paul Bunyan*. New York: Alfred A. Knopf, 1948.

60. Flint, Richard Foster. *Glacial and Pleistocene Geology*. New York: John Wiley and Sons, 1957.

61. Fouqué, F. *Santorin et Ses Éruptions*. Paris: G. Masson (ed.), Libraire de l'Académie de Médecine, 1879.

62. Fraser, George D., Eaton, Jerry P., Wentworth, Chester K. "The tsunami of March 9, 1957, on the island of Hawaii." *Bulletin of the Seismological Society of America*, vol. 49 (1959), 79-90.

63. Frazer, James George. *The Great Flood*. In *Folklore in the Old Testament*, vol. 1, chap. 4, 104-361. London: Macmillan and Co., 1919.

64. Frost, K. T. "The Critias and Minoan Crete." *Journal of Hellenic Studies*, vol. 33 (1913), 189-206.

65. Fuller, Myron L. "The New Madrid earthquake." *U.S. Geological Survey Bulletin*, no. 494 (1912).

66. Furneaux, Rupert. *Krakatoa*. Englewood Cliffs (New Jersey): Prentice-Hall, 1964.

67. Galanopoulos, Angelos G. "Die ägyptischen Plagen und der Auszug Israels aus geologischer Sicht." *Das Altertum*, vol. 10 (1964), 131-137.

68. ———. "Die Deukalionische Flut aus geologischer Sicht." *Das Altertum*, vol. 9 (1963), 3-7.

69. ———. "The Eastern Mediterranean trilogy in the Bronze Age." *Acta of the 1st International Scientific Congress on the Volcano of Thera, Greece, 1969*. Athens (1971), 184-210.

70. ———. *Greece: A Catalogue of Shocks with $I_o \geq VII$ for the Years Prior to 1800*. Athens, 1961.

71. ———. *Greece: A Catalogue of Shocks with $I_o \geq VI$ or $M \geq 5$ for the Years 1801-1958*. Athens, 1958.

72. ———. "On the location and size of Atlantis" (in Greek with English summary and captions). *Praktika tis Akademias Athenon*, vol. 35 (1960), 401-418.

73. ———. "On the origin of the Deluge of Deukalion and the myth of Atlantis." *Athenais Archaiologike Hetaireia*, vol. 3 (1960), 226-231.

74. ———. "Der Phaëthon-Mythus im Licht der Wissenschaft." *Das Altertum*, vol. 14 (1969), 158-161.

75. ———. "Tsunami. Bemerkungen zum Aufsatz 'Die Santorin-Katas-

trophe und der Exodus' von W. Krebs." *Das Altertum*, vol. 13 (1967), 19-20.

76. ———. "Tsunamis observed on the coasts of Greece from antiquity to present time." *Annali di Geofisica*, vol. 13 (1960), 369-386.

77. ———. "Zur Bestimmung des Alters der Santorin-Kaldera." *Annales Geologiques des Pays Helleniques*, vol. 9 (1958), 185-186.

78. ———, and Bacon, Edward. *Atlantis: The Truth behind the Legend*. London: Thomas Nelson and Sons, 1969.

79. Gaughran, Eugene R. L. "From superstition to science: The history of a bacterium." *Transactions of the New York Academy of Sciences*, ser. II, vol. 31 (1969), 3-24.

80. Georgalas, G. C. *Catalog of the Active Volcanoes of the World including Solfatara Fields. Part 12: Greece*. Naples: International Association of Volcanology, 1962.

81. Gilbert, G. K. In *The California Earthquake of April 18, 1906. Report of the State Earthquake Investigation Commission* (Andrew C. Lawson, Chairman). Washington, D.C.: Carnegie Institution of Washington Publication no. 87, 1908-10. vol. 1, part 1, pp. 72 and 192.

82. Goodrich, Norma Lorre. *The Ancient Myths*. New York: Mentor Books (New American Library), 1960.

83. Gordon, Cyrus H. "Notes on Minoan Linear A." *Antiquity*, vol. 31 (1957), 124-125.

84. ———. "The decipherment of Minoan." *Natural History*, vol. 72 (1963), no. 9, 22-31.

85. Gorshkov, G. S. "Gigantic eruption of the volcano Bezymianny." *Bulletin Volcanologique*, vol. 20 (1959), 76-109.

86. ———, and Dubik, Yu. M. "Directed blast on the volcano Shiveluch" ("Napravlennyy vzryv na vulkane Shiveluch"). In *Vulkany i Izverzheniya*. Moscow: Akad. Nauk SSSR, Sibirskoye Otdeleniye, Institut Vulkanologii. Izdatel'stvo "Nauka" (1969), 3-37.

87. Greer, Raymond T. "Submicron structure of 'amorphous' opal." *Nature*, vol. 224 (1969), 1199-1200.

88. Griggs, Robert C. *The Valley of Ten Thousand Smokes*. Washington, D.C.: The National Geographic Society, 1922.

89. Grimal, Pierre (ed.). *Larousse Encyclopedia of Mythology*. New York: Prometheus Press, 1959.

90. Grossling, B. F. "Seismic waves from the underground atomic explosion in Nevada." *Bulletin of the Seismological Society of America*, vol. 49 (1959), 11-32.

91. Guerber, H. A. *Myths of the Norsemen*. London: George C. Harrap and Co., 1908.

92. ———. *Myths of Northern Lands*. Cincinnati: American Book Co., 1895.

93. Gutenberg, Beno, and Richter, C. F. *Seismicity of the Earth and Associated Phenomena*. Princeton (New Jersey): Princeton University Press, 1954.

94. Hadley, Jarvis B. "The Madison landslide." *Billings [Montana] Geo-*

References

logical Society, *11th Annual Field Conference for 1960, West Yellowstone Earthquake Area,* 1960, 44-58.

95. Hallberg, Peter. *The Icelandic Saga,* trans. Paul Schach. Lincoln: University of Nebraska Press, 1962.

96. Hamilton, Edith. *Mythology.* New York: Mentor Books (New American Library), 1953.

97. Harkrider, David, and Press, Frank. "The Krakatoa air-sea waves: An example of pulse propagation in coupled systems." *Geophysical Journal of the Royal Astronomical Society,* vol. 13 (1967), 149-159.

98. Harland, J. Penrose. "Sodom and Gomorrah." In *The Biblical Archeologist Reader,* G. Ernest Wright and David Noel Freedman, eds. Chicago: Quadrangle Books, 1961, 41-75.

99. Hartland, E. S. "Mythology and folktales: Their relation and their interpretation." In *Popular Studies in Mythology, Romance and Folklore,* no. 7. London: David Nutt, 1900.

100. Hatai, Shinkishi. "The earthquake as one of the determining factors of the organismal equilibrium in nature." *Proceedings of the Fifth Pacific Science Congress* (Canada, 1933), vol. 2 (1934), 915-931.

101. ———, and Abe, N. "The responses of the catfish, *Parasilurus asotus,* to earthquakes." *Proceedings of the Imperial Academy* (Tokyo), vol. 8 (1932), 375-378.

102. ———, Kokubo, S., and Abe, N. "The earth currents in relation to the response of catfish." *Proceedings of the Imperial Academy* (Tokyo), vol. 8 (1932), 478-481.

103. Healy, James. "Geological observations on Thera." *Acta of the 1st International Scientific Congress on the Volcano of Thera, Greece, 1969.* Athens, 1971, 180-183.

104. ———. "The tempo of rhyolitic volcanism in New Zealand." *Acta of the 1st International Scientific Congress on the Volcano of Thera, Greece, 1969.* Athens, 1971, 64-72.

105. Hédevári, Péter. "Energetical calculations concerning the Minoan eruption of Santorini." *Acta of the 1st International Scientific Congress on the Volcano of Thera, Greece, 1969.* Athens, 1971, 257-276.

106. Heezen, Bruce C. "A time clock for history." *Saturday Review,* December 6, 1969, 87-90.

107. Hennig, Richard. "Altgriechische Sagengestalten als Personifikation von Erdfeuern und vulkanischen Vorgangen." *Deutsches Archäologisches Institut Jahrbuch,* Band 54 (1939), 230-246.

108. Herbert, Don, and Bardossi, Fulvio. *Kilauea: A Case History of a Volcano.* New York: Harper and Row, 1968.

109. Herz, O. F. "Frozen mammoth in Siberia" (Extracts translated from the Russian). Smithsonian Institution Annual Report for 1903, 611-625.

110. Hobbs, William Herbert. *Earthquakes: An Introduction to Seismic Geology.* New York: D. Appleton and Co., 1907.

111. Hodgson, John H. *Earthquakes and Earth Structure.* Englewood Cliffs (New Jersey): Prentice-Hall, 1964.

112. Hofsten, Nils von. "Olaus Rudbeck." In *Swedish Men of Science,* Sten Lindroth, ed. Stockholm: The Swedish Institute/ Almqvist and Wiksell, 1952.

113. Hood, Sinclair. "Archeology in Greece, 1961-62." [British School in Athens] *Archeological Reports for 1961-62,* 3-31.

114. ———. *The Home of the Heroes: The Aegean before the Greeks.* London: Thames and Hudson, 1967.

115. ———. "The International Scientific Congress on the Volcano of Thera, 15th-23rd September 1969." *Kadmos,* vol. 9 (1970), 98-106.

116. ———. "Late Bronze Age destructions of Knossos." *Acta of the 1st International Scientific Congress on the Volcano of Thera, Greece, 1969.* Athens, 1971, 377-383.

117. ———. *The Minoans: Crete in the Bronze Age.* London: Thames and Hudson, 1971.

118. Hutchinson, R. W. *Prehistoric Crete.* Baltimore: Penguin Books, 1962.

119. Hwang, Li-San, and Lin, Albert C. "Experimental investigations of wave run-up under the influence of local geometry." *Proceedings of the International Symposium on Tsunamis and Tsunami Research, University of Hawaii, 1969.* Honolulu, 1970, 406-425.

120. Iacopi, Robert. *Earthquake Country.* Menlo Park (California): Lane Book Co., 1964.

121. Jaggar, T. A. *Volcanoes Declare War: Logistics and Strategy of Pacific Volcano Science.* Honolulu: Paradise of the Pacific, Ltd., 1945.

122. Jahn, Melvin E., and Woolf, Daniel J. (translators and annotators). *The Lying Stones of Dr. Johann Bartholomew Adam Beringer, Being His Lithographiae Wirceburgensis.* Berkeley: University of California Press, 1963.

123. Keller, Jörg. "Datierung der Obsidiane und Bimstoffe von Lipari." *Neues Jahrbuch für Geologie und Paläontologie, Monatshefte,* 1970, no. 2, 90-101.

124. ———. "The major volcanic events in recent eastern Mediterranean volcanism and their bearing on the problem of the Santorini ash layers." *Acta of the 1st International Scientific Congress on the Volcano of Thera, Greece, 1969.* Athens, 1971, 152-169.

125. ———. Personal communication, 1972.

126. Keller, Werner. *The Bible as History: Archeology Confirms the Book of Books.* London: Hodder and Stoughton, 1956.

127. Kelly, Allan O., and Dachille, Frank. *Target: Earth.* Pensacola (Florida), 1953.

128. Kramer, Samuel Noah. *Cradle of Civilization.* New York: Time, Inc., 1967.

129. Kranz, Walter. "Vulkanexplosionen, Sprengtechnik, praktische Geologie und Ballistik." *Zeitschrift der Deutschen Geologischen Gesellschaft,* vol. 80 (1928), 257-307.

References

130. Krebs, Walter. "Die Santorin-Katastrophe und der Exodus." *Das Altertum*, vol. 12 (1966), 135-144.

131. Kretikos, N. A. "Sur des phénomènes sismiques produits avant et depuis l'éruption du volcan du Santorin." *Comptes Rendus Hebdomédaires des Séances de l'Académie des Sciences* (Paris), tome 181 (1925), 923-926.

132. ——— [Critikos, N. A.]. "Sur la séismicité des Cyclades et de la Crète." *Annales de l'Observatoire national d'Athènes*, vol. 9 (1926), (Reviewed by I. Friedlander in *Zeitschrift für Vulkanologie*, vol. 10 [1927], 219).

133. Kunz, George Frederick. *The Curious Lore of Precious Stones.* Philadelphia: J. B. Lippincott, 1913; New York: Dover Publications, 1971. (Page references in the text are to the 1971 edition.)

134. Kvale, Anders. "Recent crustal movements in Norway." *Annales Academiae Scientarum Fennicae (Suomalaisen Tiedeakatemian Toimituksia)*, ser. A-III, no. 90 (1966), 213-221.

135. Lamb, H. H. "Volcanic dust in the atmosphere; with a chronology and assessment of its meteorological significance." *Philosophical Transactions of the Royal Society of London*, sec. A, vol. 266, no. 1178 (1970), 425-533.

136. Landen, David. "Alaska earthquake, 24 March 1964." *Science*, vol. 145 (1964), 74-76.

137. Lear, John. "The volcano that shaped the western world." *Saturday Review*, November 5, 1966, 57-60 and 63-66.

138. Lebedinsky, V. I. *Volcanoes and Man* (Vulkany i Chelovek). Moscow: Izdatel'stvo "Nedra," 1967.

139. Leet, L. Don. *Causes of Catastrophe.* New York: McGraw-Hill, 1948.

140. Ley, Willy. *Another Look at Atlantis, and Fifteen Other Essays.* Garden City (New York): Doubleday and Company, 1969, 1-15.

141. Libby, Willard F. *Radiocarbon Dating.* Chicago: University of Chicago Press, 1955.

142. Livingstone, D. C. "Certain topographical features of northeastern Oregon and their relation to faulting." *Journal of Geology*, vol. 36 (1928), 694-708.

143. Luce, J. V. *Lost Atlantis: New Light on an Old Legend.* New York: McGraw-Hill, 1969. Published in England under the title *The End of Atlantis: New Light on an Old Legend.* London: Thames and Hudson, 1969.

144. Mackenzie, Donald A. *Myths of China and Japan.* London: Gresham Publishing Company, 1923.

145. Marinatos, Sp[yridon]. "Amnisos, die Hafenstadt des Minos." *Forschungen und Fortschritte*, vol. 10 (1934), 341-343.

146. ———. *Crete and Mycenae.* New York: Harry N. Abrams, Inc., 1960.

147. ———. "La 'diaspora' créto-mycénienne." In Les peuples de l'Europe du Sud-Est et leur rôle dans l'histoire (Antiquité). *Association internationale d'Études du Sud-est Européen, Rapport pour la Séance*

Plénière. Premier Congrès International des Études Balkanique et Sud-est Européennes, Sofia, 1966, 60-74.

148. ———. "The volcanic destruction of Minoan Crete." *Antiquity,* vol. 13 (1939), 425-439.

149. ———. "Some words about the legend of Atlantis." *Archeiologicon Deltion* no. 12 (Athens), 1969.

150. Marinos, G., and Melidonis, N. "About the size of the sea-wave (tsunami) during the prehistoric eruption of Santorin" [in Greek]. *Bulletin of the Geological Society of Greece (Hellenikes Geologikes Hetairias),* vol. 4 (1959-61), 210-218.

151. ———, and ———. "On the strength of seaquakes (tsunamis) during the prehistoric eruptions of Santorini." *Acta of the 1st International Scientific Congress on the Volcano of Thera, Greece, 1969.* Athens, 1971, 277-282.

152. Marriott, Alice Lee. "Beowulf in South Dakota." *New Yorker,* August 2, 1952, 46-51.

153. Marshall, Patrick. *Geology of Mangaia.* Honolulu, Hawaii: Bernice P. Bishop Museum, 1927.

154. Mattison, Ray H. *Devil's Tower National Monument—A History.* Devil's Tower Natural History Association, 1967.

155. Matz, Friedrich. *Minoan Civilization: Maturity and Zenith.* Cambridge Ancient History, rev. ed., fasc. 12 (chaps. IV(b) and XII), 1962 (p. 46).

156. Mavor, James W., Jr. *Voyage to Atlantis.* New York: G. Putnam's Sons, 1969.

157. McWilliams, Carey. "The folklore of earthquakes." *American Mercury,* vol. 29 (1933), June, 199-201.

158. Menyaylov, I. A., Nikitina, L. P., Khramova, G. G. "Gas-hydrothermal eruption of the volcano Ebeko in February-April 1967" ("Gazogidrotermal'noye izverzheniye vulkana Ebeko v fevrale-aprele 1967 g."). *Akad. Nauk SSSR, Sibirskoye Otdeleniye, Institut Vulkanologii, Byulleten' Vulkanologicheskikh Stantsiy,* no. 45 (1969), 3-6.

159. Michael, Henry N., and Ralph, Elizabeth K. "Discussion of radiocarbon dates obtained from precisely dated sequoia and bristlecone pine samples." *International Radiocarbon Conference in New Zealand, October 18-25, 1972,* preprint.

160. Milanovskiy, Ye. Ye. "Atlantis in the Aegean Sea?" ("Atlantida v Egeyskom more?"). *Priroda,* 1960, no. 1, 3-6.

161. Miller, Don J. "The Alaska earthquake of July 10, 1958: Giant wave in Lituya Bay." *Bulletin of the Seismological Society of America,* vol. 50 (1960), 253-266.

162. Moore, Carleton B. Oral communication, 1970.

163. Moore, James G. "Petrology of deep-sea basalt near Hawaii." *American Journal of Science,* vol. 263 (1965), 40-52.

164. ———. "Base surge in recent volcanic eruptions." *Bulletin Volcanologique,* vol. 30 (1967), 337-363.

References

165. ————, and Richter, Donald H. "Lava tree molds of the September 1961 eruption, Kilauea Volcano, Hawaii." *Bulletin of the Geological Society of America*, vol. 73 (1962), 1153-1158.

166. Myres, John Linton. *Who Were the Greeks?* New York: Biblo and Tanner, 1967.

167. Neumann van Padang, M. "Two catastrophic eruptions in Indonesia, comparable with the plinian outburst of the volcano of Thera (Santorini) in Minoan time." *Acta of the 1st International Scientific Congress on the Volcano of Thera, Greece, 1969.* Athens, 1971, 51-63.

168. Ninkovich, Dragoslav. Personal communication (April 14, 1972).

169. ————, and Heezen, Bruce C. "Physical and chemical properties of volcanic glass shards from the pozzuolana ash, Thera Island, and from upper and lower ash layers in eastern Mediterranean deep-sea sediments." *Nature*, vol. 213 (1967), 582-584.

170. ———— and ————. "Santorini tephra." *Colston* [Research Society] *Papers*, vol. 17 (1965), 413-453.

171. Odell, N. E. "Mount Ruapehu, New Zealand: Observations on its crater lake and glaciers." *Journal of Glaciology*, vol. 2 (1965), 601-605.

172. Oostdam, B. L. "Age of lava flows on Haleakala, Maui, Hawaii." Bulletin of the Geological Society of America, vol. 76 (1965), 393-394.

173. Ouwehand, Cornelis. *Namazu-e and Their Themes: An Interpretative Approach to Some Aspects of Japanese Folk Religion.* Leiden: E. J. Brill, 1964.

174. Page, D[enys] L. *The Santorini Volcano and the Destruction of Minoan Crete.* London: The Society for the Promotion of Hellenic Studies, 1970.

175. ————. "The volcano at Santorini and the problem of Minoan Crete: An introduction to the historical and archaeological problem." *Acta of the 1st International Scientific Congress on the Volcano of Thera, Greece, 1969.* Athens, 1971, 371-376.

176. Parasnis-Carayannis, George. "A study of the source mechanism of the Alaska earthquake and tsunami of March 27, 1964." *Pacific Science*, vol. 21 (1967), 301-310.

177. Pariyskiy, N. N., Artamasova, G. N., and Kramer, M. V. "On the problem of the role of tidal stresses as the trigger mechanism in earthquakes" ("K voprosu o roli prilivnykh napryazheniy kak spuskovogo mekhanizma pri zemletryaseniy"). In *Fizicheskiye Osnovaniya Poiskov Metodov Prognoze Zemletryaseniy.* Moscow: Akad. Nauk SSSR Institut Fiziki Zemli, 1970, 62-63.

178. Parker, K. Langloh. *Australian Legendary Tales.* New York: The Viking Press, 1966.

179. Partsch, Joseph. "Geologie und Mythologie in Kleinasien." *Philologische Abhandlungen* (Berlin: Martin Hertz), 1888; 105-122.

180. Pendlebury, J. D. S. *A Handbook to the Palace of Minos—Knossos.* London: Max Parrish and Company, 1954.

181. Petersen, Ole V. Oral communication, 1970.

182. Plato. The *Timaeus* and the *Critias*.
  a) In The Loeb Classical Library. London: William Heinemann, vol. 7, pp. 1-307, 1929.
  b) In B. Jowett. *The Dialogues of Plato*. London: Oxford University Press, 437-543, 1892.
  c) *The Timaeus and the Critias*, or *Atlanticus*. Thomas Taylor translation. New York: Pantheon Books, 1944.

183. Platon, Nicholas. *Zakros: The Discovery of a Lost Palace of Ancient Crete*. New York, Charles Scribners Sons, 1971.

184. Polach, Henry. Oral communication, 1971.

185. Pollitt, Jerome J. "Atlantis and Minoan civilization: An archeological nexus." *Yale Alumni Magazine*, vol. 33 (1970), no. 5, 20-28.

186. Pomerance, Leon. "The final collapse of Santorini (Thera). 1400 B.C. or 1200 B.C.?" *Acta of the 1st International Scientific Congress on the Volcano of Thera, Greece, 1969*. Athens, 1971, 384-394.

187. ———. "The final collapse of Santorini (Thera): 1400 B.C. or 1200 B.C.?" *Studies in Mediterranean Archeology*, vol. XXVI (1970). Göteborg: Paul Åströms Förlag.

188. Poole, Lynn, and Poole, Gray. *Volcanoes in Action: Science and Legend*. New York: McGraw-Hill, 1962.

189. Pope, Maurice. "Aegean writing and Linear A." *Studies in Mediterranean Archeology*, vol. VIII. Lund (Sweden), 1964.

190. Popham, Mervyn. "Late Minoan pottery, a summary." *Annual of the British School at Athens*, no. 62 (1967), 337-351.

191. Press, Frank, and Harkrider, David. "Air-sea waves from the explosion of Krakatoa." *Science*, vol. 154 (1966), 1325-1327.

192. Prins, J. E. "Characteristics of waves generated by a local surface disturbance." *University of California Institute of Engineering Research, Wave Research Laboratory*, Berkeley, ser. 99, issue 1, 1956.

193. Ramskou, Thorkild. "Solstenen." *Skalk* (1967), no. 2, 16-17.

194. Reck, Hans. *Santorin: Der Werdegang eines Inselvulkans und sein Ausbruch 1925-1928. Ergebnisse eine Deutsch-Griechischen Arbeitsgemeinschaft. Band I. Die Geologie der Ring-Inseln und der Kaldera von Santorin*. Berlin: Andrews and Steiner, 1936.

195. Reed, A. W. *Aboriginal Fables and Legendary Tales*. Sydney-Wellington-Auckland: A. H. and A. W. Reed, 1965.

196. ———. *Legends of Rotorua and the Hot Lakes*. Wellington (New Zealand): A. H. and A. W. Reed, 1958.

197. ———. *Myths and Legends of Australia*. Sydney-Wellington-Auckland: A. H. and A. W. Reed.

198. ———. *A Treasury of Maori Folklore*. Wellington-Auckland-Sydney: A. H. and A. W. Reed, 1963.

199. Rekstad, J., and Vogt, J. H. L. "Søndre Helgelands Kvartaergeologi." In "Praktisk-geologisk undersøgelser af Nordlands amt, pt. III. Søndre Helgeland." Norges Geologisk Undersøgelse, 1900, 95-105.

References

200. Renault, Mary. *The Bull from the Sea*. New York: Pantheon Books, 1962.

201. ———. *The King Must Die*. New York: Pantheon Books, 1959.

202. Richard, J. J., and Neumann van Padang, M. *Catalog of the Active Volcanoes of the World including Solfatara Fields, part 4: Africa and the Red Sea*. Naples, International Volcanological Association, 1957.

203. Rickard, T. A. *Man and Metals: A History of Mining in Relation to the Development of Civilization*. New York: McGraw-Hill (Whittlesey House), 1932. In two volumes.

204. Rikitake, T. "Detectability of seismomagnetic effect." *EOS* (Transactions of the American Geophysical Union), vol. 50 (1969), p. 399.

205. ———. "A resistivity variometer." *EOS* (Transactions of the American Geophysical Union), vol. 50 (1969), p. 400.

206. Rittman, Alfred. *Volcanoes and Their Activity*, trans. E. A. Vincent. New York-London: John Wiley and Sons, 1962.

207. Robinson, Herbert Spencer, and Wilson, Knox. *The Encyclopaedia of Myths and Legends of All Nations*. London: Edmund Ward, 1962.

208. Rogers, J. E. T. *Bible Folk-Lore, A Study in Comparative Mythology*. London: Kegan Paul, Trench and Co., 1884.

209. Rogers, Robert William. *Cuneiform Parallels to the Old Testament*. New York: Eaton and Mains; Cincinnati: Jennings and Graham.

210. Rose, H. J. *A Handbook of Greek Mythology, Including Its Extension to Rome*. London: Methuen and Co.

211. ———. *Gods and Heroes of the Greeks: An Introduction to Greek Mythology*. Cleveland and New York: The World Publishing Company (Meridian Books), 1958.

212. Rothé, Peter, and Schmincke, Hans-Ulrich. "Contrasting origins of the eastern and western islands of the Canarian archipelago." *Nature*, vol. 218 (1968), 1152-1154.

213. Schoo, Jan. *Hercules' Labors, Fact or Fiction?* Chicago: Argonaut, Inc., 1969.

214. ———. "Vulkanische und seismische Aktivität des Ägaischen Meeresbeckens im Spiegel der Griechischen Mythologie." *Mnemosyne* (Bibliotheca Classica Batava), ser. 3, vol. 4 (1936-37), 257-294.

215. Sheridan, Michael F. Personal communication, 1970.

216. Simkin, Tom, and Howard, Keith A. "Caldera collapse in the Galapagos Islands, 1968." *Science*, vol. 169 (1970), 429-437.

217. Smith, Homer William. *Man and His Gods*. New York: Grossert and Dunlap, 1952.

218. Snell, Leonard J. "Effect of sedimentation on ancient cities of the Aegean coast, Turkey." *Bulletin of the International Association of Scientific Hydrology*, année 8 (1963), no. 4, 71-73.

219. Solov'yev, S. L. "The tsunami problem and its significance for Kamchatka and the Kurile Islands" (Problema tsunami i yeye znacheniye dlya Kamchatki i Kuril'skikh ostrovov). In *Problema Tsunami*. Moscow: Izdatel'stvo "Nauka," 1968, 7-50.

220. Spanuth, Jürgen. *Das enträtselte Atlantis.* Stuttgart: Union Deutsch. Verlagsgesellschaft, 1953.

221. Spence, Lewis. *The History of Atlantis.* New Hyde Park (New York): University Books, 1968.

222. ———. *The Problem of Atlantis.* New York: Brentano's, 1925.

223. Stearns, Harold T. *Geology of the State of Hawaii.* Palo Alto (California): Pacific Books, 1966.

224. ———. *Road Guide to Points of Geologic Interest in the Hawaiian Islands.* Palo Alto (California): Pacific Books, 1966.

225. Steinbrugge, K. V., and Moran, D. F. "Damage caused by the earthquakes of July 6 and August 23, 1954." *Bulletin of the Seismological Society of America,* vol. 46 (1956), 15-33.

226. ———, and ———. "Engineering aspects of the Dixie Valley-Fairview Peak earthquakes." *Bulletin of the Seismological Society of America,* vol. 47 (1957), 335-352.

227. Stewart, J. A. *The Myths of Plato.* London: Centaur Press, 1960.

228. Stuiver, Minze, and Suess, Hans E. "On the relationship between radiocarbon dates and the true sample age." *American Journal of Science, Radiocarbon Supplement,* vol. 8 (1966), 534-540.

229. Suess, Eduard. *The Face of the Earth.* Oxford: Clarendon Press, 1904 (vol. 1, part 1, chapter 1).

230. Suryakanta, Shastri. *The Flood Legend in Sanskrit Literature.* Delhi: S. Chand and Co., 1950.

231. Swire, Otha F. *Skye: The Island and Its Legends.* London: Oxford University Press, 1952.

232. Takeuchi, Hitoshi; Uyeda, Seiya; and Kanamori, Hiroo. *Debate about the Earth: Approach to Geophysics through Analysis of Continental Drift,* trans. Keiko Kanamori. San Francisco: Freeman, Cooper, 1967.

233. Tamrazyan, G. P. "Tide-forming forces and earthquakes." *Icarus,* vol. 7 (1967), 59-65.

234. Termier, Pierre. "Atlantis." *Annual Report of the Smithsonian Institution,* 1915, 219-234.

235. Thorarinsson, Sigurdur. *Askja on Fire.* Reykjavik: Almenna Bókafélagid, 1963.

236. ———. "Damage caused by tephra falls in some big Icelandic eruptions and its relation to the thickness of the tephra layer." *Acta of the 1st International Scientific Congress on the Volcano of Thera, Greece, 1969.* Athens, 1971, 213-236.

237. ———. *"The eruptions of Hekla in historical times: A tephrochronological study"* (Part 1 of *The Eruption of Hekla,* 1947-1948). Reykjavík: Vísindefélag Íslendinga, 1967.

238. ———. *Hekla on Fire.* Munich: Hans Reich Verlag, 1956.

239. ———. *Hekla: A Notorious Volcano.* Reykjavík: Almenna Bókafélagid, 1970.

240. ———. "The Lakagígar eruption of 1783." *Bulletin Volcanologique,* vol. 33 (1969), 910-929.

References

241. ———. Personal communication, 1970.

242. ———. *Surtsey: The New Island in the North Atlantic.* Reykjavík: Almenna Bókafélagid.

243. ———, Einarsson, T., and Kjartansson, G. "On the geology and geomorphology of Iceland." *Geografiska Annaler* (1959), 135–169.

244. Thornbury, William D. *Principles of Geomorphology.* New York: John Wiley and Sons, 1965.

245. ———. *Regional Geomorphology of the United States.* New York: John Wiley and Sons, 1965.

246. Thucydides. *The Peloponnesian War,* trans. Crawley. New York: Random House (Modern Library), 1951.

247. Tocher, Don. "The Dixie Valley-Fairview Peak, Nevada, earthquakes of December 16, 1954. Introduction." *Bulletin of the Seismological Society of America,* vol. 47 (1957), 299-300.

248. Tsuya, H. (ed.) "The Fukui earthquake of June 28, 1948." *Report of the Special Committee for the Study of the Fukui Earthquake.* Tokyo, 1950, pp. 26-27 and 156-157.

249. Umbreit, Wayne. *Modern Microbiology.* San Francisco: W. H. Freeman and Co., 1962.

250. Vaughan, Agnes Carr. *The House of the Double Axe: The Palace at Knossos.* Garden City (New York): Doubleday and Company, 1959.

251. Velikovsky, Immanuel. *Ages in Chaos,* vol. I: From the Exodus to King Akhnaton. Garden City (New York): Doubleday and Company, 1952.

252. ———. *Earth in Upheaval.* New York: Dell Publishing Company, 1968.

253. ———. *Worlds in Collision.* Garden City (New York): Doubleday and Company, 1950; New York: Dell Publishing Company, 1967. (Page references in the text are to the 1967 edition.)

254. Verbeek, R. D. M. *Krakatau.* Batavia, Imprimérie de l'État, 1886.

255. Vermeule, Emily. "The fall of Knossos and the Palace Style." *American Journal of Archeology,* vol. 67 (1967), 195-199.

256. ———. *Greece in the Bronze Age.* Chicago: University of Chicago Press, 1964.

257. ———. "The promise of Thera: A Bronze Age Pompeii." *Atlantic Monthly,* vol. 220 (1967), 83-94.

258. Vigfusson, Gudbrand, and Powell, F. York (editors and translators). *Cristne Saga, The Story of the Conversion in Iceland.* In *Origines Islandicae.* Oxford: Clarendon Press, 1905, 370-418.

259. Vine, F. J., and Matthews, D. H. "Magnetic anomalies over oceanic ridges." *Nature,* vol. 199 (1963), 947-949.

260. Vink, B. W., and Schuiling, R. D. "Estimates of the various types of energy released by the eruption of Thera, Greece, at about 1400 B.C." *Acta of the 1st International Scientific Congress on the Volcano of Thera, Greece, 1969.* Athens, 1971, 288-290.

261. Vitaliano, Charles J., and Vitaliano, Dorothy B. 1973. "Volcanic tephra on Crete." In press.

262. Vitaliano, Dorothy B. "Atlantis: A review essay." *Journal of the Folklore Institute* (Indiana University), vol. 8 (1971), 66-76.

263. ———. "Bemerkungen zu A. G. Galanopoulos, 'Der Phaethon-Mythus im Licht der Wissenschaft.'" *Das Altertum*, vol. 16 (1970), 82-83.

264. ———. "Geomythology: The impact of geologic events on history and legend, with special reference to Atlantis." *Journal of the Folklore Institute* (Indiana University), vol. 5 (1968), 5-30.

265. ———, and Vitaliano, Charles J. "Plinian eruptions, earthquakes, and Santorin—A review." *Acta of the 1st International Scientific Congress on the Volcano of Thera, Greece, 1969.* Athens, 1971, 88-108.

266. Washington, Henry S. "Santorini eruption of 1925." *Bulletin of the Geological Society of America*, vol. 37 (1925), 349-384.

267. Weiner, J. S. *The Piltdown Forgery*. London: Oxford University Press, 1955.

268. Werner, Alice. *Myths and Legends of the Bantu*. London: George G. Harrap and Co., 1933.

269. Werner, E. T. C. *Myths and Legends of China*. London: George G. Harrap and Co., 1922.

270. Westervelt, W. D. *Hawaiian Legends of Volcanoes*. Rutland (Vermont) and Tokyo: Charles E. Tuttle Company, 1963.

271. Wilcox, Ray E. "Some effects of recent ash falls with especial reference to Alaska." *U.S. Geological Survey Bulletin* 1028-N (1959), 409-476.

272. Williams, Howel. *The Ancient Volcanoes of Oregon*. Eugene: Oregon State System of Higher Education, 1953.

273. ———. "Calderas and their origin." *University of California Publications, Bulletin of the Department of Geological Sciences*, vol. 25, 239-346.

274. ———. *Crater Lake: The Story of Its Origin*. Berkeley: University of California Press, 1941.

275. ———. Personal communication, 1972.

276. Woolley, Charles Leonard. *Excavations at Ur: A Record of Twelve Years' Work*. London: Ernest Benn, 1954.

277. Zarudski, Edward F. K. "Geophysical study of Santorini (Thera)." *Acta of the 1st International Scientific Congress on the Volcano of Thera, Greece, 1969.* Athens, 1971, p. 351.

INDEX

aa flow, 23, 113
aborigines of Australia, 60-63, 74-75
Aci Trezza, 139-40
Adams, Mount, 50-52
Aegean Sea, 157-58, 179-217, 231-51, 273
African legends: on earthquakes, 82-83; on floods, 164-65
Akrotiri, 205, 207, 211
Alaska: earthquakes, 12, 101, 193; glacial advance and retreat, 30-32; legends on glaciation, 30-32, 35
Algonquin Indians, 81
Alkmena, 270
alluvial fan, 63
Almannagjá, 21
Aloipuaa's rock, 116
Alpheus, 65
Althing, 20-21
Amenhotep IV, 253
American Indian lore: absorption of later elements by, 169; and Beowulf legend, 151-52; flood stories, 169-72; on Hell's Canyon area, 54-56; on mountains, 50-56; on volcanoes, 123-27
Amnisos, 216
Anaphi, 249
andalusite, 79-80
Andree, Richard, 153, 158-59, 166-67, 172, 176
Annals of the Bishops, 128
Ansei era, 83
Antillia, 223
Apache flood legend, 170-71
Apollo moon samples, 74
Apollodorus, 246
Aqaba, Gulf of, 263-64
Araucanian Indians, 81, 173-74
archeology, Minoan, 184-217
Arethusa, 65
Argive plain, 246
Argonauts, 76, 249
Aristotle, 86, 94, 158
Arizona flood legends, 170-72
Armstrong, B. H., 102-3
Artemis, 65

Ásbyrgi depression, 41-42
ash, volcanic: described, 107n; from Santorin, 184-85, 203-4, 256-58; and hail, 258; and plagues, 256-57; *see also* pumice; tephra
Asian flood legends, 161-64
Askja, 192
Aspronisi, 181
Assam, 85
asthenosphere, 226
Athena, 246
Atlantic Ocean, 14, 218-29
Atlantis, 218-51, 273
Atlantis: The Antediluvian World, 221
Atlas, 49-50, 86, 235
Atlas Mountains, 49-50
Atna Indians, 32
Augustine, 259
Australia: flood traditions, 165-66; legends on minerals, 74-75; legends on rivers and lakes, 60-63
australites, 72-74
Aztec traditions, 123, 175-76

Ba Ouvando, 82
Babylonian flood tradition, 153-58
Bacon, Edward, 263, 270
Bacon, Sir Francis, 221
Baiame, 62
Balos, 205
Banská Štiavnica, 75
base surge, 189
Basoga, 85-86
Basques, 230
Batok, 135-36
bediasites, 73
Bellerophon, 67, 246-47
Ben Gurion, David, 263
Benioff, Hugo, 227
Bennett, J. G., 254, 260
Benua-Jakun tribe, 164
Beowulf legend, 151-52
Berezovka mammoth, 279-81
Berg, L. S., 231
Bering Strait, 28-30
Beringer, J. B. A., 6, 275-76
Bezymianny, 12, 193

Index

delta, 63
deluge lore, 142-78
Dettifoss waterfall, 19
Demavend, 138
Deukalion, 156-59, 232, 246, 259-61, 274
Devil's Tower, 41-44
Diamond Head, 106-7
diluvium, 143
Diodorus Siculus, 247
Dixie Valley, Nev., 13, 101
Dolomieu, Déodat de, 140
Dolphin Ridge, 222-24
domes, volcanic, 38, 48-49
Donnelly, Ignatius T. T., 221-222
Dorson, Richard M., xi, 1, 56
Dover cliffs, 39
drainage patterns, 61-63
Drebkhuls, 82
Dry Falls, Wash., 145-46
Du-mu, 163-64
dwarfs, 67-68

earth: geology as science of, 2; internal constitution of, 92
"earthquake weather," 94
earthquakes: causes of, 91-95; classification of, 91-93; cracks or fissures from, 95-101; Cretan, 183-217; dangers of, 101; folklore on, 81-103; and Santorin eruption, 194-96; subjectivity in reporting, 12-13; submarine, and tsunamis, 148-50; tectonic, 93, 194-96; volcanic, 93, 194-96
Ebeko volcano, 210n
Edo earthquake, 84
Egmont, Mount, 119-20
Egypt: chronology of, 198; lack of flood traditions in, 164-65; Santorin eruption's effect on, 252-66
Ehrenberg, C. G., 255
Eldhraun, 17
Ellis, William, 109-11, 117, 167
entrenched stream, 60
Ephesus, 32, 34
erosion along a joint, 40
eruptions: as distinct from "explosions," 188n; of Santorin, 179-217; see also volcanoes
estuary, defined, 53
Etna, Mount, 139
euhemerism, 1, 8
Euhemerus, 1, 4
Euphrates River, 154-56
Eusebius, 259
eustatic fluctuations, 28

Evans, Sir Arthur, 196, 202
Exodus, 254-66, 274

"fairy stones," 69-70
fakelore, 1, 56-58
Fallon, Nev., 12-13
Fara, 154
faulting, 91
Fiji Islands, 167
fimbulvetr, 35, 267
fires, natural earth, 67
fissures, 95-101
flank eruptions, 109
flash floods, 144-45
flashing arcs, 268-69
Flatey Book, 130
floating islands, 270
flood lore, 142-78
flotation process, 76
folklore: defined, 3; modern, 56-58; relationship to geology, 1-2; see also myths
fossil shells, 229
Fountain of Arethusa, 65
frauds, 6-7, 275-77
Frazer, Sir James, 153-55, 159-60, 163, 169, 247
Frisian legend, 39
frogs, plague of, 255-56
Frost, K. T., 231
frozen mammoths, 279-81
Fuego, 195
Fujiyama (Fuji-san), 131-33
Fukui earthquake, 98-99
fulgurites, 73
fumarolic activity, 210n-11n
Funza River, 173, 175

gabbro, 47-48
Galanopoulos, A. G., xi, 158, 232-37, 251, 254-55, 260-63, 267-70
Gannet Island, 122
Gediz River, 32-33
Gefion, 66-67
gems, 68; see also minerals
Genesis, Noah in, 153-54
geology: fast vs. slow processes of, 8-10, 27; and paleoclimatology, 228-29; relationship to folklore, 1-2; time scale of, 9
geomythology: conception of term, xi; defined, 1, 3; see also mythology
German giant myths, 37-39
giant lore, 37-50
Giant's Causeway, 43, 45
Gilbert, G. K., 96-98

299

Index

Index

Thera, 180-81, 194, 205-7, 231, 250; *see also* Santorin, Kalliste
Therasia, 181, 205
Theseus, 244-46
Thingvellir, Iceland, 21-22
Thom, King, 253, 260
Thorarinsson, S., 128
Three Sisters, 50
Thucydides, 196, 244
tidal waves, 86 n
Tigris River, 154-56
Timaeus, 218-19, 235, 246, 248, 251, 267
time scale, geologic, 9
Titans, 49
Tlascaltans, 82
Tofua, 85
Tongans, 85
Tonganui, 59
Tongariro, 119, 122
Tonopah, Nev., 75
Torghatten, 39-41
translations, faulty, 5-6
tree molds, 114-15
Tristan da Cunha, 13-15, 48
Troezen, 246
Troy, 32-33, 35
Tsimshian Indian, 32
tsunamis: Biblical references to, 265; defined, 86 n; described, 148-50; danger from, 100; effect on Exodus? 262-64; and flooding, 146, 159-60, 174; height of, 149-50, 193; propagation of, 148; from Krakatau, 185, 187; from Santorin, 192-94, 250, 273
tuff, 107
tuffs, welded, 170-71
Tuthmosis III, 240, 260
twilight compass, 77
twin crystals, 69-70
Tyre, 32

Ugarit, 271
Umbreit, W., 255
Ur, 154, 156
Usu, 48

Utnapishtim, 153-54, 156

van Bemmelen, R. W., 25-26
Van Helmont, J. B., 87
Vapheio cups, 244
Velikovsky, Immanuel, 165, 177, 260
Ventris, Michael, 201
Vikings' compass, 77-80
Virunga volcanoes, 135, 137
Visu, 133
volcanic domes, 38, 48-49
volcano lore, 104-41
volcanoes: active, definition of, 53; and earthquakes, 194-96; newsworthiness of, 11-12; Santorin, 179-217
Vulcan, 141

Wanyamwasi, 83
Washington *Post*, 12-13
Washington state: Indian lore on landforms, 53-56; mountain mythology, 50-52
water: on earth, 144; ground, 65-66; scarcity and vanishing civilizations, ⁓34; source in springs, 65-66
waves: higher than tsunamis, 193 n; highest, 193 n; seismic, 91-92, 187, 194, 225 n; *see also* tsunamis
Wegener, Alfred, 225
Weiner, J. S., 276
welded tuffs, 170-71
Welsh flood legend, 160
White Cliffs of Dover, 39
White Island, New Zealand, 121-22
Wizard Island, 58
Woolf, D. J., 275
Worlds in Collision, 165
Wyeast, 50
Wyoming's Devil's Tower, 41-44

Yakutat Bay area, Alaska, 30-32

Zaire, volcanoes in, 135-38
Zakros, 216
Zephaniah, 264-65
Zeus, 49, 270

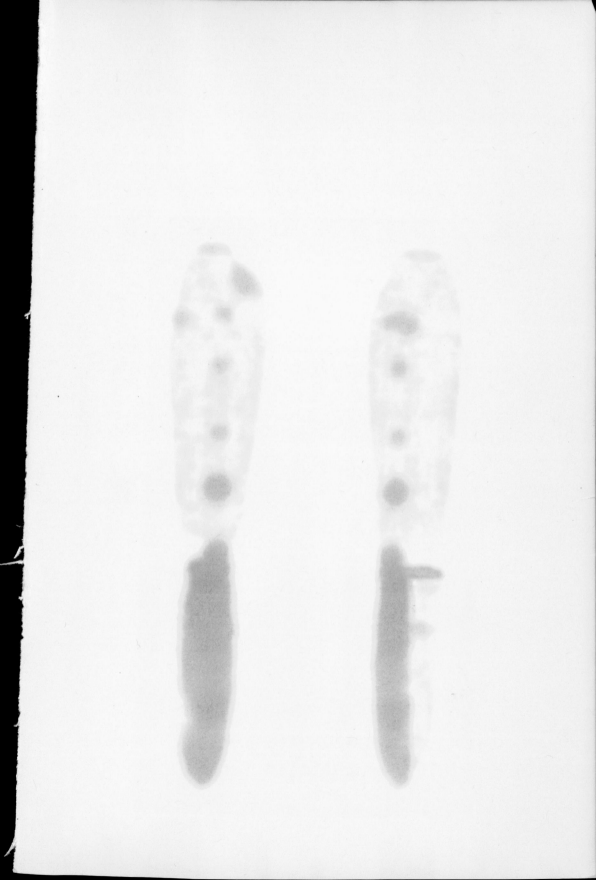

DATE DUE

1/26 3³⁰ pm	
1/26 5³⁰ pm	
2/23 10:00am	

GAYLORD   PRINTED IN U.S.A.